FOREWORD

OECD Member countries, in signing the United Nations Framework Convention on Climate Change, have recognised that the industrialised countries must take responsibility for past greenhouse gas emissions. They have further taken the lead in making climate change policy commitments. Most of them are already in the process of developing their individual national plans to meet these commitments.

The core of the work of the International Energy Agency is reviewing and debating the national energy policies of our Member countries. In recent years much of our attention has focussed on integrating into energy policy the serious energy-related environmental problems, of which climate change is one of the most threatening. From this, we know that the solutions for meeting commitments to stabilise or reduce greenhouse gas emissions from energy sources must be tailored to each country's circumstances. Solutions must also be tested against criteria, such as comparability, cost-effectiveness, and effects on trade and energy security. The sort of "peer review" which IEA conducts of its Members' energy policies is helping to accomplish this and to find comparable and feasible national policies.

This publication focusses exclusively on greenhouse gas emissions from energy sources and the commitments that countries are taking towards reducing their emissions of these gases. Each Member country has contributed to and endorsed the national country profile included for it. Thus, this publication is an authoritative source of information on the commitments by OECD Member countries and the factors influencing these countries in establishing their positions on their commitments. In addition, much of the growth in emissions will nevertheless come from energy sources outside OECD Member countries. Hence, the publication includes the national positions and the factors influencing them for those countries outside of OECD which are either key emitters of greenhouse gases themselves many of which were key actors in the negotiations of the Framework Convention on Climate Change.

This report is published under my responsibility as Executive Director of the IEA and does not necessarily reflect the views or policies of the IEA or its Member governments.

Helga Steeg
Executive Director

CLIMATE CHANGE
POLICY INITIATIVES

INTERNATIONAL ENERGY AGENCY

2, RUE ANDRÉ-PASCAL, 75775 PARIS CEDEX 16, FRANCE

The International Energy Agency (IEA) is an autonomous body which was established in November 1974 within the framework of the Organisation for Economic Co-operation and Development (OECD) to implement an international energy programme.

It carries out a comprehensive programme of energy co-operation among twenty-two* of the OECD's twenty-four Member countries. The basic aims of the IEA are:

i) co-operation among IEA participating countries to reduce excessive dependence on oil through energy conservation, development of alternative energy sources and energy research and development;

ii) an information system on the international oil market as well as consultation with oil companies;

iii) co-operation with oil producing and other oil consuming countries with a view to developing a stable international energy trade as well as the rational management and use of world energy resources in the interest of all countries;

iv) a plan to prepare participating countries against the risk of a major disruption of oil supplies and to share available oil in the event of an emergency.

* IEA participating countries are: Australia, Austria, Belgium, Canada, Denmark, Finland, France, Germany, Greece, Ireland, Italy, Japan, Luxembourg, the Netherlands, New Zealand, Norway, Portugal, Spain, Sweden, Switzerland, Turkey, the United Kingdom, the United States. The Commission of the European Communities takes part in the work of the IEA.

ORGANISATION FOR ECONOMIC CO-OPERATION AND DEVELOPMENT

Pursuant to Article 1 of the Convention signed in Paris on 14th December 1960, and which came into force on 30th September 1961, the Organisation for Economic Co-operation and Development (OECD) shall promote policies designed:

— to achieve the highest sustainable economic growth and employment and a rising standard of living in Member countries, while maintaining financial stability, and thus to contribute to the development of the world economy;

— to contribute to sound economic expansion in Member as well as non-member countries in the process of economic development; and

— to contribute to the expansion of world trade on a multilateral, non-discriminatory basis in accordance with international obligations.

The original Member countries of the OECD are Austria, Belgium, Canada, Denmark, France, Germany, Greece, Iceland, Ireland, Italy, Luxembourg, the Netherlands, Norway, Portugal, Spain, Sweden, Switzerland, Turkey, the United Kingdom and the United States. The following countries became Members subsequently through accession at the dates indicated hereafter: Japan (28th April 1964), Finland (28th January 1969), Australia (7th June 1971) and New Zealand (29th May 1973). The Commission of the European Communities takes part in the work of the OECD (Article 13 of the OECD Convention). Yugoslavia has a special status at OECD (agreement of 28th October 1961).

CLIMATE CHANGE POLICY INITIATIVES

TABLE OF CONTENTS

SECTION 3   NON-OECD MEMBER COUNTRY DESCRIPTIONS

**SECTION 4   UNITED NATIONS FRAMEWORK CONVENTION ON
                          CLIMATE CHANGE**

GLOSSARY

TABLES

FIGURES

INTRODUCTION

For the duration of the negotiations to develop a framework convention on climate change (roughly from the beginning of 1991 through the middle of 1992), the IEA periodically produced and updated information on the energy-related climate change policy initiatives of the members of OECD. Late in 1991, eleven other countries which were key to the negotiations and which have large national energy-related CO_2 emissions were added to the collection of country profiles. The country profiles and key energy and CO_2 emissions data and other indicators were then re-issued on a periodic basis as the countries' positions changed or firmed up or new initiatives were announced. The information was produced in loose-leaf format and contained in a binder to facilitate the updating.

This publication provides the positions of countries as they enter the post-negotiation phase. Herein are fully revised, concise descriptions of the energy-related climate policy initiatives by the 24 OECD Member countries, and 13 non-OECD countries which have the largest energy-related CO_2 emissions. In this publication three profiles (Kazakhstan, the Russian Federation and Ukraine) replace the former USSR which was formally dissolved on 1 January 1992 in the midst of negotiations on the framework convention.

These profiles include statements and announcements made at the United Nations Conference on Environment and Development held in Rio de Janeiro from 3-14 June 1992 as well as factors influencing countries' decisions. The key energy and CO_2 emissions data and derivative indicators have all been updated to 1990, the base year chosen by the majority of OECD Member countries for establishing their CO_2 or greenhouse gas emissions targets.

SECTION 1

OVERVIEW OF CLIMATE CHANGE POLICY INITIATIVES

1. THE FRAMEWORK CONVENTION ON CLIMATE CHANGE —
   A GLOBAL COMMITMENT

Intense international negotiations in the sessions of the United Nations Intergovernmental Negotiating Committee (INC) produced the Framework Convention on Climate Change (hereafter Convention) which in May 1992 was adopted by the INC. The adopted text, attached as Section 4, is still considered an advance copy, i.e. awaiting the legal designation of "satisfied" text which can only be achieved when there is full agreement of text in all official United Nations' languages. In total 154 countries and the European Community, including all OECD Member countries except Turkey, signed the Convention at the United Nations Conference on the Environment and Development (hereafter UNCED). For the Convention to go into effect, at least 50 countries must ratify it, at which point they become Parties to the Convention. For purposes of understanding its significance as a commitment by Parties, the most important aspects are summarised here.

In general, it can be said that the Convention represents commitments to an overall global objective, guiding principles, actions and institutions. The objective of the Convention is "stabilization of greenhouse gas concentrations in the atmosphere at a level that would prevent dangerous anthropogenic interference with the climate system". Principles were ultimately included after long debate over wording. They are important for the operative concepts which are woven into them, such as:

- equity; present and future generations; sustainable development;
- common but differentiated responsibilities and respective capabilities; developed countries take the lead;

- full consideration of specific needs and special circumstances of developing countries;
- precautionary measures; lack of full scientific certainty is not a reason for postponing action;
- measures should be cost-effective and comprehensive, cover all relevant sources, sinks and reservoirs and adaptation, and comprise all economic sectors;
- policies should be appropriate and integrated, taking into account economic growth; and
- in maintaining an open international economic system, measures taken should not constitute a restriction on international trade.

The commitments contained in the Convention are stated in the form of legally binding obligations. Those to be undertaken by all Parties to the Convention are:
- preparation of national inventories of greenhouse gas emissions caused by human activity using comparable methodologies;
- development of programmes to mitigate effects of greenhouse gases and measures of adaptation to climate change;
- co-operation on technology related to greenhouse gas emissions for all relevant sectors;
- sustainable management of greenhouse gas sinks and reservoirs;
- integration of climate change considerations with other policies;
- research to reduce uncertainties concerning scientific knowledge of climate change, the effects of the phenomenon and the effectiveness of responses to it;
- exchange of information, including on technology and on the economic consequences of actions covered by the convention;
- education, training and public awareness raising; and
- communication of information on implementation.

Co-ordination of economic and administrative instruments is called for as is the identification of policies which would increase emissions above those that would otherwise occur. Flexibility was given in implementation of commitments to economies in transition (i.e. from command to market economies). Full consideration was called for in funding, insurance and technology transfer to especially adversely affected countries, for example small island countries and least developed countries. Consideration also will be given to countries with economies which are highly dependent on income generated from the production, processing and export, and/or consumption of fossil fuels and associated energy-intensive products and/or the use of fossil fuels where there is serious difficulty in switching.

Institutions created are the Conference of the Parties and two subsidiary bodies. The primary functions of the Conference of the Parties will be:
- promotion and facilitation of exchange of information on measures adopted by Parties;
- co-ordination of measures adopted;

- development and refinement of comparable methodologies, inter alia, for inventories of greenhouse gas emissions and for evaluating the effectiveness of measures; and

- assessment of the overall effects of the measures taken, in particular environmental, economical and social effects and cumulative impacts.

The subsidiary bodies created are one for Scientific and Technological Advice and one for Implementation. The first provides information and advice on scientific and technological matters (*inter alia,* innovative, efficient and state-of-the-art technologies, international co-operation in research and development, and questions on scientific, technological and methodological issues). It will comprise government representatives and draw on existing international bodies. The implementation subsidiary body will assess and review the effective implementation of the Convention including the reporting done under it.

Prompt Start. Because ratification by 50 countries could take as long as two years, negotiators devised an interim means for continuing intergovernmental work on climate change. At the last session of the INC a resolution was adopted allowing action after the signing but before the entry into force of the Convention in order to, *inter alia,* prepare for the first session of the Conference of the Parties and for the items which the Convention requires to be included on its agenda. The resolution in effect continues the INC as it was constituted with its *ad hoc* secretariat and allows it to have additional sessions as necessary until the first session of the Conference of the Parties and the designation of the secretariat of the Convention by the Conference of the Parties. The resolution "invites" States to communicate information regarding measures relevant to the provisions of the Convention. At UNCED many OECD Member countries re-iterated the need for a prompt start and the President of the United States called for the first post-UNCED meeting on climate change to be held before 1 January 1993. The Chancellor of Germany offered to host the first Conference of the Parties.

2. NATIONAL COMMITMENTS TO CLIMATE CHANGE POLICIES AND THE CONVENTION

In the negotiations, three types of specific national commitments to be taken only by Developed Country Parties were discussed and ultimately resolved in the Convention: stabilizing or reducing greenhouse gas emissions, financing the incremental costs imposed on developing country parties by the Convention and facilitating technology transfer or co-operation. The first was the primary preoccupation of the OECD Member countries while the last two were promoted more by the developing countries. The next four sub-sections discuss these commitments further.

Emissions Stabilization, Targets and Complementary Concepts. While no concrete targets with timetables for greenhouse gas emissions stabilization were included in the Convention, developed countries agreed in essence to adopt national policies and take measures consistent with the objective of returning their anthropogenic emissions of CO_2 and other greenhouse gases to 1990 levels by the end of this decade. The Convention calls for a review of the adequacy of this and other specific commitments by developed countries at the first session of the Conference of the Parties (and regularly thereafter). Several OECD countries openly stated at UNCED their dissatisfaction with this unspecific formulation on emissions in the Convention. A few (e.g. Austria, the Netherlands and Switzerland) attempted but failed to obtain a separate declaration on stabilization of national CO_2 emissions at 1990 levels to be launched at UNCED. In light of this, it is quite likely that the Conference of the Parties will revisit this commitment early on.

With respect to emissions stabilization, throughout the negotiations there was strong, almost unanimous, support among OECD Member countries to establish emissions targets. Table 1 sets out the status as of UNCED of these national commitments. Most of the targets were announced early in, or prior to the start of, the negotiations although modifications and clarifications were made up to and even after the last INC session. For example, the United Kingdom announced just before UNCED a more ambitious target of stabilizing its CO_2 emissions by 2000 rather than the year 2005. The target would still be conditional on other countries taking similar actions.

Box 1 characterises the approach to commitments by OECD countries. This box illustrates the diversity of approaches taken under the general rubric of "target". A few countries have set unilateral targets but many recognise the need for other countries to take similar commitments and make this a condition to achieving their own target. Several European countries still have no individual target commitment but were included under the European Community Council's Luxembourg agreement of 29 October 1990 which set a goal of stabilization of overall EC emissions of CO_2 at 1990 levels by the year 2000. Since then, a few have designated their own targets, e.g. Spain and Luxembourg. European Free Trade Association (EFTA) members in turn agreed in November 1990 to adopt the same overall CO_2 target as the European Community although most already had their own targets by this point. Only one country, Spain, has a target adjusted for economic growth, though it might be assumed that Greece and Portugal will eventually adopt economic growth-adjusted targets. The European Community's own regional stabilization target is designed to balance out their growth in emissions by reductions in emissions by other EC Members. Two countries (France and Japan), adopted per-capita stabilization targets (which allows for some growth in emissions). The United States alone adopted a set of policies to reduce the growth in emissions. Differences among countries also occur in the base years chosen, although 1990 predominates, and on whether to stabilize or reduce CO_2 alone or all greenhouse gases.

Outside the OECD group, few countries have made public statements about emissions stabilization. Notably, of the largest emitters outside OECD, only Poland had stated that its emissions would be no higher than 1988-89 levels in the year 2000. Hungary made a similar statement at UNCED for different base years.

Figure 1 shows total OECD and non-OECD projected emissions from 1990 to 2000 and an estimate of the effects of the OECD commitments made to date. Together these undertakings, if fully achieved, would account for slightly less than 3% reduction of estimated global emissions of CO_2 in the year 2000. That is, instead of growing approximately 22% in this time period, global emissions would only grow about 19%. Clearly, even the ambitious commitments by OECD Member countries will be dwarfed by growth in emissions in the large, rapidly growing developing countries. Furthermore, as reported in Box 1, only a handful of OECD Member countries have funded action plans and/or carbon taxes in place to support their commitments. Even those countries with carbon or CO_2 taxes are still predicting that their greenhouse gas emissions in the year 2000 will exceed stabilization. This is illustrated for a number of OECD Member countries in Figures 2a and 2b which show actual emissions of CO_2 in 1990, the projected emissions for the year 2000 and the target adopted by that country for that year. These countries were chosen for illustration here because they were reviewed "in-depth" by the IEA in its normal energy policy review process in the 1991 cycle and adequate data were available. The projections of energy supply and demand upon which the emissions estimates are based will be published in *Energy Policies of IEA Countries, 1991 Review* (forthcoming IEA, 1992).

Joint Implementation. A few countries, in analyzing their forecasted emissions vis-a-vis their announced targets have acknowledged that meeting them might be harder (or more costly) than originally expected. Two of these countries (Norway and Germany) started investigations into the idea of investing in emissions reduction projects in other countries where such projects would be more cost-effective than trying to achieve an equivalent reduction within their own countries. This is a derivative of the idea of emissions trading and has been called variously "clearinghouse for projects", "joint implementation" or "compensation measures". The common theme is achieving greater cost-effectiveness than could be achieved by solely investing in projects within national boundaries given that there are large differences in the opportunities for and national costs of response strategies. (Hereafter, the reference will be to "joint implementation" despite the slight differences in the concepts being proposed.)

Norway originated the idea of joint implementation as a system of credit towards the achievement of a national emissions target for a country investing in emissions reductions projects elsewhere (outside national boundaries). In discussions with other countries, Norway emphasized the need to develop a mechanism to implement the concept. Germany in its 9 December 1991 "Second Report of the 'Energy Supply' Subgroup in the Interministerial Working Group on 'CO_2 Reduction'" of the German Federal Ministry of Economics found that its target (see Table 1) would be difficult or impossible to reach without supplementary measures such as, *inter alia,* joint implementation. The concept again is that compensatory measures taken (including the creation of sinks) in one facility or country could be allowed to offset emissions in another.

Provision for joint implementation is allowed in the language of the Convention although full development of the concept is delegated to the Conference of Parties at its first meeting. The idea of joint implementation should be kept distinct from the joint target shared by countries which are members of the European Community. In this case, as laid out in Box 1, more aggressive targets in some countries would offset the economic-growth adjusted targets of others, such as that adopted by Spain. It is entirely possible that there could be joint implementation within the European Community as well as a joint target.

CO$_2$ and Carbon Taxes. Until recently, energy taxation generally has been used to raise government revenue. Energy or environmental policy considerations have not been a driving force behind the fiscal structure of energy products, with the exception of tax differentials between gasoline and diesel, and between leaded and unleaded gasoline. In many IEA countries, a tax advantage was given to diesel fuels after the first oil shock, in recognition of the higher energy efficiency of diesel engines.

Many IEA countries are rationalising and restructuring their fiscal policies. This is affecting the balance of taxes applied to different energy products and consumer categories. At the same time, the use of taxes to influence consumer behaviour and to internalise external costs, particularly environmental costs, is attracting growing attention from policy makers. In some countries, environmental taxes or charges (related to pollutant emission levels) on energy products or activities finance pollution control or energy efficiency measures, notably in the Netherlands and France. In the Netherlands, where the burden of environmental charges on energy products is among the highest in the OECD, taxes are used for revenue raising, particularly to transfer funds from the private sector to the public sector for the financing of environmental protection measures such as monitoring and enforcing regulations. In France, a tax on SO_2, NO_x and hydrochloric acid emissions is applied to all industries with combustion facilities of more than 200 MW. Revenue from the tax is used to finance air quality improvement and monitoring projects, including grants for the installation of pollution control equipment.

Over the last two years, carbon or CO_2 taxes have been enacted in Denmark, Finland, the Netherlands, Norway and Sweden, starting with Finland in 1990. Table 2 provides details on these countries' carbon taxes. As noted in the table, the carbon or CO_2 taxes have been variously applied, i.e. with simultaneous reform of energy taxes (Sweden), with other environmental taxes (Netherlands, Finland), as a mixed tax on energy content and on carbon content (Denmark and in a later reform of its carbon/environmental tax, the Netherlands). Norway and the Netherlands increased their carbon taxes in 1992 and Sweden reduced its carbon tax for industry and raised it for residential users while simultaneously eliminating the energy tax for industry, beginning in 1993.

These CO_2 and related taxes have to be seen in the context of the sizeable taxes such as excise tax and VAT already imposed on fuels for fiscal or other reasons not necessarily related to the fuels' environmental effects. Thus the impact of carbon taxes tends to be moderated. Generally, the net price effect of the taxes cannot be expected to influence behaviour significantly. In addition, significant exemptions to the new taxes are extended everywhere except the Netherlands, to various categories of energy users, such as coal users, energy-intensive industries, electricity generators or international air or sea travel. In Sweden, though the carbon tax itself is relatively large, its effect is limited by the fact that its introduction was offset by substantial reductions in other energy taxes. The nominal rate on coal is significant in Sweden, but with exemptions granted for industry and power generation and a virtual absence of coal consumption in the non-industrial sectors, the amount of tax collected on coal use is very small. Given the exemptions and adjustments in various countries, the taxes cannot be said to be strictly based on the carbon content of fuels.

Other OECD countries and the European Community are considering a range of policy instruments, including carbon taxes, but most are reluctant to introduce such taxes unilaterally because of uncertainty about the effects and concerns that their international competitiveness might suffer. They are aware that models used to evaluate the impact of such policies are imprecise and that results of such modelling must therefore be considered in the context of the assumptions used. Different models produce widely different projections of CO_2 emissions from region to region. Even the same model can generate significantly different estimates on relatively small differences in assumptions about exogenous factors such as population growth rates or the rate of autonomous energy efficiency improvements.

Furthermore, there is a wide range of estimates of the GDP losses that would occur by around 2050 if the OECD countries made sizeable cuts in their CO_2 emissions. There are also indications that during an initial adjustment period, the costs (including social adjustment costs) could be large owing to rapid down-sizing of energy-intensive industries; the international competitiveness of some energy-intensive sectors and energy-producing countries could be significantly reduced. The GDP costs of reducing CO_2 emissions would likely be less if measures were phased-in rather than being introduced at once. Studies and modelling of carbon taxes by the IEA and the OECD Economics Department (using the "GREEN" model) have also shown that:

- it would take substantially different levels of tax between regions and countries to generate the same percentage of reduction in CO_2 emissions in each;
- to continue to be effective, carbon taxes have to increase with time;
- taxes imposed unilaterally or even regionally would have little effect on the level of world emissions — much less on long-term concentrations of CO_2 in the atmosphere — mostly because of the very strong economic growth and resulting growth in emissions taking place in countries outside the OECD that, as yet, do not plan to stabilize or reduce emissions, and possibly also because of the inevitable flight of carbon-intensive industry and production to regions or countries without such taxes;
- the availability of technologies and opportunities for fuel substitution are key; without non-fossil fuel options, the upper limit on the tax required rises, and the cost and availability of low-carbon or carbon-free technology is critical in limiting the level required; and
- eliminating energy subsidies (the vast majority of which exist outside OECD countries) would result in a significant reduction in emissions and yield net economic benefits, especially outside the OECD.

Funding for Climate Change Policies and Programmes Outside of OECD. Additional funding for climate change studies and their eventual implementation was a subject of intense debate within the climate negotiations from the first meeting. In return for developing countries acceding to the general commitments laid out for all countries signing the Convention, developed countries agreed to including commitments to new and additional funds and greater technology co-operation for developing countries. Specifically OECD Members and the European Commission are bound by the Convention to:

- "provide new and additional financial resources to meet the agreed full costs" of compliance by developing countries, resources (including technology transfer) for agreed full incremental costs for implementing covered measures and assistance to particularly vulnerable countries for adaptation; and

- promote, facilitate and finance the transfer of environmentally sound technologies.

The financial mechanism will provide for financial resources on a grant or concessional basis, including for the transfer of technology. Its Secretariat functions would be in the interim carried out by the General Assembly of the United Nations in resolution 45/212 (that is, continuation of present Secretariat). The Global Environment Facility (GEF) will be entrusted on an interim basis with the operation of the financial mechanism. The GEF is jointly operated by the World Bank, the United Nations Environment Programme and the United Nations Development Programme and funds environmental projects related to, *inter alia,* climate change.

Several countries, such as Denmark, Norway, Switzerland as well as the European Community, announced rather early in the negotiations that they would favour additional funding and would be able to contribute although no figures on the amounts were given. The United States "broke the ice" on announcing specific amounts of additional funding when it announced on 28 February 1992 during a negotiating session that it would give $25 million for analysis of national climate change strategies as well as $50 million more to the GEF. At UNCED the Prime Minister of Britain announced substantial extra funds for, *inter alia,* energy efficiency in its aid programme and most OECD countries explicitly promised replenishment of the GEF.

The issue of funding took on a much wider significance at UNCED than it had during the negotiation of the Convention. At UNCED, the "Agenda 21", which covers all environment and development problems, including protection of the atmosphere and the subset of climate change issues, was adopted. While it is non-binding, it had been estimated by the United Nations Secretariat that Agenda 21 would cost about US$600 billion annually, between now and the year 2000. Various estimates of the amount of new commitments to fund all of Agenda 21 have been made. They range from US$2.5 to 3 billion over the next several years. Clearly, only part of this would go to fund climate change-related actions.

Technology Co-operation. The issue of transfer of technology was hotly debated throughout the negotiations on the Convention and continued to be very difficult to resolve at UNCED. For the Convention, negotiators finally agreed upon financing for the transfer of technologies as mentioned above. Also mentioned above all parties to the Convention agreed to exchange of information on technologies (an important pre-requisite to technology transfer) and the subsidiary body on Scientific and Technological Advice will also become involved in technology transfer. Furthermore, the Article on Research and Systematic Observation of the Convention allows for support of international and intergovernmental efforts" to strengthen ... and to promote access to, and the exchange of data and analyses ..." for carrying out commitments.

Technology transfer and co-operation was also a major issue at UNCED. The final version of Agenda 21 includes Chapter 34, "Transfer of Environmentally Sound Technology,

Cooperation and Capacity-Building," which covers the technological needs for the full range of environment and development issues contained in Agenda 21. At UNCED many countries (both developed and developing) referred to the importance of resolving the issue of technology transfer to the satisfaction of all parties. More specifically, in his speech to UNCED, the Prime Minister of the United Kingdom announced a new global Technology Partnership initiative to promote technology co-operation in achieving sustainable development. It will be launched by a Technology Partnership Conference in the first half of 1993. These commitments at UNCED would serve the broader set of environment and development problems but would include climate change.

To meet needs of countries for information on technologies relevant to climate change policies, the IEA and the OECD are setting up the IEA/OECD Technology Information Exchange (TIE). The objective of TIE is to facilitate the diffusion and exchange of technical data and economic information with regard to the technology options and measures that can be adopted in order to control, stabilize and mitigate greenhouse gas emissions. It will accomplish this by:

- establishing a comprehensive directory of technology information sources and expertise in the area of mitigation of greenhouse gases, compatible with needs and capabilities of the target customer group; and

- creating a network connecting national and multilateral clearinghouses and information centres in the area of greenhouse gases mitigation, in order to improve information exchange amongst participants.

TIE was presented at UNCED in Rio de Janeiro and is presently being adapted to future client needs, as expressed in responses to an IEA/OECD questionnaire.

3. KEY COMPARATIVE ENERGY AND CO_2 EMISSIONS DATA

Table 3 provides Key Energy and CO_2 Emissions Data for OECD Member countries relevant to climate change policies [1]. Table 4 parallels the data provided in Table 3 but for the top thirteen CO_2-emitting non-OECD Member countries. Table 5 provides the relevant percentages of the energy-related CO_2 emissions for the top 25 energy-related CO_2-emitting countries globally. Table 6 illustrates the breakdown of CO_2 emissions based on IEA statistics for all the Commonwealth of Independent States (i.e. ex-USSR Republics). This affects the relative ranking of the countries of the world in terms of energy-related CO_2 emissions. The dissolution of the USSR resulted in three of the republics (Kazakhstan, Russia and Ukraine) being placed among the top CO_2-emitters. Russia ties with China for second highest in the world.

---

1. CO_2 accounts for the largest share of radiative forcing due to increased greenhouse gas emissions, but other important contributors are methane, chlorofluorocarbons and nitrous oxide. All greenhouse gases and both their sources and sinks must be accounted for in measuring contributions to net greenhouse gases. However, limitations of IEA data restrict further discussion to CO_2 emissions only.

CO_2-Emissions Estimating Methodology and the Importance of Biomass. The estimates of CO_2 emissions in Tables 3, 4 and 5 are calculated using the IPCC/OECD methodology which is fully described in the IPCC's Estimation of Greenhouse Gas Emissions and Sinks, Background Report from the OECD Experts' Meeting, 18-21 February 1991. While the Convention left it to the Conference of the Parties to agree on the methodology for emissions inventories to be used for the purposes of the Convention, it is generally felt that the IPCC/OECD methodology will be adopted.

A follow-up IPCC workshop on 5-6 December 1991 in Geneva resolved a number of outstanding issues on the methodology, including treatment of biomass and international marine bunkers. Marine bunkers are now included and biomass excluded in the Adjusted TPES and resulting CO_2 emissions estimates reported in all the tables and at the end of each country profile in Sections 2 and 3. Peat is treated as non-renewable and is included in the CO_2 estimates for each country. Because this is the first time the figures are reported with these adjustments, Tables 3 and 4 also show the breakdown for CO_2 emissions for both marine bunkers and biomass (less peat) for each country.

Even though this methodology has been adopted by the IPCC/OECD working groups, their treatment of biomass still cannot handle some of the problems encountered with treating all "other solid fuels" (except peat) as biomass. This is because "other solid fuels" (besides peat) includes wood, wood waste, vegetal waste and black liquor which are possibly sustainably produced as well as industrial waste, municipal waste and other non-specified solid fuels which are a mixture of possibly sustainable and not sustainable components.

It will still be important in the future to obtain further precision. Some examples from OECD illustrate why. For the case of Sweden, about 92% of the CO_2 from combustion of "other solid fuels" could possibly be sustainably produced and therefore considered as recyclable CO_2 and is properly subtracted from Sweden's total CO_2. This amount is significant as it represents just over 10% of Sweden's total energy-related CO_2 emissions. For Greece, "other solid fuels" is reported to the IEA as virtually all wood combustion and, if sustainably produced, would represent 100% recyclable carbon from "other solid fuels" combustion or slightly over 2% of its total CO_2 emissions. Ireland's "other solid fuels" are reported to be virtually all peat, representing 100% non-recyclable CO_2. Again, using the IPCC/OECD methodology which counts peat as non-renewable is correct. Alternatively, Switzerland's "other solid fuels" are a mixture of wood, woodwaste, municipal and industrial waste of which only approximately 35% is possibly sustainable and the rest of which is a mixture of biomass and other components.

Because the IEA energy balances and many countries' own statistics are not complete at this level of disaggregation for "other solid fuels", the IPCC/OECD working group found it necessary to use the simplifying assumption that all "other solid fuels" (except peat) are sustainably-produced biomass. Even where the data are available, emission factors for some of these fuels are highly uncertain or unavailable. So a "biomass" emission factor was adopted. However, until the IPCC/OECD working group succeeds in completing full emissions inventories for all greenhouse gases and for all sources and sinks, it is likely that some CO_2 will not be accounted for by the present approach to biomass treatment. This would occur in those cases where the biomass is not sustainably produced.

Finally, the exception to Norway's carbon tax on coal (exempting the non-energy uses of coal) brings out another area of uncertainty with estimation of CO_2 emissions. The problem occurs because the emission factors for the CO_2 emissions from coal used in industrial processes such as metallurgy are not defined. This would be a minor portion of most countries' CO_2 emissions but for Norway around 88% of its coal use is in such metallurgical processes. The OECD is working with the IPCC to resolve some of the outstanding issues with greenhouse gas emissions, beginning with CO_2, and this problem will be one of those examined.

Historical and Projected Energy-Related CO_2 Emission Trends. Figure 1 focusses on a small slice of the future (1990 to 2000) which is the horizon of the majority of the commitments contained in OECD Member countries' emissions targets (some have targets which extend to the year 2005). Figure 3[2] looks at the same division between OECD Member countries and other countries showing the historical evolution of actual (estimated) CO_2 emissions as well as future emissions through the year 2000. This figure illustrates that in about the year 1985 there was a crossover of total emissions stemming from OECD as a whole as compared to those from the rest of the world when OECD emissions became the smaller portion.

Figure 3 also illustrates how CO_2 emission growth rates within OECD as a whole had actually fallen dramatically (from a high in 1976 to a low in 1980) and then began to rise again afterward but much more slowly. Within this period, annual CO_2 emissions from the energy sector in the OECD varied considerably and between 1977 and 1983 actually declined roughly 1 200 million metric tons of CO_2. During those years the OECD experienced major increases in oil prices, a four-fold increase in the contribution of nuclear energy and the attendant reduction of oil in electricity generation, along with a significant recession. This provides a sense of the relative magnitude of potential economic, structural and other changes that might be implied by interventions to reduce greenhouse gas emissions over the next ten years such as those represented by the commitments already made by OECD Member countries. This is why it is important to structure carefully any interventions contemplated. A major difference could be made in the transition costs by making the revenues of taxes imposed as "revenue-neutral" as possible and finding ways to maximise the benefit of investment in new equipment and infrastructure occurring and contained within and outside OECD Member countries, thus achieving offsetting positive economic effects.

---

2. For both Figures 3 and 4, sources of data include:
   - for 1925 to 1959 for all countries and for 1960 to 1970 for non-OECD countries, *Energy in the World Economy,* by Darmstatter, Teitelbaum, and Polach, John Hopkins University Press, 1971
   - for 1960 to 1990 for OECD countries, IEA databases built from country submissions
   - for 1971 to 1990 for non-OECD countries, IEA databases
   - for 1991 to 2000, for OECD countries, IEA databases based on country submissions of forecasts, for non-OECD, the IEA World Energy Outlook

It should be noted that eastern Germany is included as OECD (for all years).

It should also be noted that non-commercial fuels are generally those which are not bought or sold, i.e. they are bartered or gathered. Until 1959 for OECD and until 1970 for non-OECD, some non-commercial fuels might have been included in the consumption data. After those dates, non-commercial fuels are not included. There are some inconsistencies because the category, "other solid fuels" can include traded combustible materials such as wood waste and municipal solid waste and some OECD countries may be reporting some of these fuels while other countries do not.

Figure 4 provides cumulative CO_2 emissions, both historical and prospective. This helps to illustrate the growing role of CO_2 emissions for countries not belonging to the OECD. Up to 1990 the OECD accounted for approximately 61% of cumulative CO_2 emissions. The cumulative emissions between 1990 and 2000 from the rest of the world could be roughly one and a quarter times those of OECD Member countries for the same ten-year time period. This shows the growing importance of finding means for spurring the most cost-effective measures to reduce CO_2 emissions and other greenhouse gases outside as well as inside OECD Member countries.

Key Contributing Factors and Emissions Patterns of Non-OECD Countries. Section 3 contains country profiles of Brazil, China, the Czech and Slovak Federal Republic, India, Kazakhstan, Mexico, Poland, Republic of South Korea, Romania, the Russian Federation, Saudi Arabia, South Africa and Ukraine. These profiles contain country positions, factors and key data parallel to those for OECD Member countries in Section 2. These 13 countries are the largest energy-related CO_2-emitters outside OECD and represent 39% of present global (and 75% of non-OECD) emissions of energy-related CO_2. For three of the countries (China, Poland and South Africa) coal represents over 70% of TPES and for two others (Czech and Slovak Federal Republic and India) over 50%. Within the TPES represented by commercial fuel use, fossil fuel dependency is over 90% for all but two (Brazil and the Republic of South Korea) of the 13, and most of them have few affordable fuel-switching opportunities.

For countries outside of the OECD region, there are two distinct groups: those with expectations for rapid industrialisation, population growth, urbanisation and increasing standard of living, implying an increasing share of energy and hence emissions over the years and those which are already industrialised but restructuring their economies. Most developing countries have a lot of sub-optimum combustion equipment while economies in transition may be operating obsolete equipment. In both cases this decreases efficiency of energy use. Furthermore, Brazil, China and India use significant amounts of non-commercial fuels. For Brazil, non-commercial fuels use represents over 50% addition to TPES of commercial fuels. For China it is less than 10% more while for India it is one-third more. A wide variety of fuels burnt at very low efficiencies are included in the category, "non commercial fuels". The conversion factors used in reporting their consumption, expressed in heat value (Mtoe), do not take account of the efficiency of fuel use. The continuing trend to move away from non-commercial fuels has meant increasing use of fossil fuels although generally in more efficient technologies. But CO_2 emissions, as accounted for in the IPCC/OECD methodology would, however, increase — because CO_2 from biomass is presently subtracted from CO_2 from energy emissions inventories.

Finally, several of these countries are also major exporters of fossil fuels. For example, in ranking of the largest world hard coal exporters, South Africa is third largest; the USSR was fourth largest (exports now split between Kazakhstan, Russia and Ukraine); Poland is sixth largest; and China is seventh largest. Similarly for oil exports, Kazakhstan, Mexico, Russia and Saudi Arabia are all major world oil exporters.

Many of these thirteen non-OECD countries (particularly Brazil, China, India and Saudi Arabia, but also Russia taking over for the former USSR) have been very influential in the IPCC and INC processes. Furthermore, China hosted, and a number attended, the Beijing Ministerial Declaration on Environment and Development (see excerpts at the end of Section 3).

Box 1

**Characterisation of Approach to Commitments
by OECD Member Countries**

Unilateral, unconditional commitments to targets (Austria, Canada, Iceland, Luxembourg and Switzerland) of which two elaborated by funded plans of action and supported by carbon or CO_2 taxes (Netherlands and Denmark);

Unilateral, but preliminary non-binding and/or conditional commitments to targets (Australia, Germany, Italy, New Zealand and the United Kingdom) of which two supported by carbon tax (Finland and Norway);

Target adjusted for need for economic growth (Spain);

Conditional targets based on per capita emissions (France and Japan);

Commitment to a set of policies which will stabilize emissions (United States);

Regional targets balanced out by allowing economic-growth-adjusted targets of some countries to be offset by the more aggressive targets of other countries (the European Community);

Targets not specified but implicit in membership in European Community or EFTA (Belgium, Greece, Iceland, Ireland and Portugal) of which one supported by carbon tax (Sweden).

Table 1: **Status of Commitments of OECD Countries on Global Climate Change**

Country	Type of Commitment	Gases Included	Action	Base Year	Commitment Year	Conditions/Comments
Australia	Target	NMP GHG	Stabilization 20% Reduction	1988 1988	2000 2005	Interim planning target; to be implemented if others take like action.
Austria*	Target	CO_2	20% Reduction	1988	2005	Still needs parliamentary approval.
Belgium**	EC Agreement	CO_2	(see footnote)			—
Canada	Target	CO_2 and other GHG	Stabilization	1990	2000	CFCs will be phased out by 1997, methyl chloroform by 2000, and other major ozone-depleting substances by 2005.
Denmark**	Target	CO_2	20% Reduction	1988	2005	Implementation plan adopted.
Finland*	Target	CO_2	Stabilization	1990	2000	Policy goal, not a formal target.
France**	Target	CO_2	Stabilization	1990	2000	This is a per capita per year target of less than 2 metric tons of carbon
Germany**	Target	CO_2	25-30% Reduction	1987	2005	—
Greece**	EC Agreement	CO_2	(see footnote)			—
Iceland*	EFTA Agreement	CO_2	(see footnote)			—
Ireland**	EC Agreement	CO_2	(see footnote)			—
Italy**	Target	CO_2	Stabilization 20% Reduction	1988 1988	2000 2005	Non-binding resolution.
Japan	Target	CO_2	Stabilization	1990	2000	. on per capita basis; . implemented if others act likewise.
Luxembourg**	Target	CO_2	Stabilization 20% Reduction	1990 1990	2000 2005	—

Table 1 (Continued): **Status of Commitments of OECD Countries on Global Climate Change**

Country	Type of Commitment	Gases Included	Action	Base Year	Commitment Year	Conditions/Comments
Netherlands**	Target	CO_2	Stabilization 3-5% Reduction	89/90 89/90	1995 2000	Unilateral action committed
	Target	All GHG	20-25% Reduction	89/90	2000	Unilateral action committed.
New Zealand	Target	CO_2	20% Reduction	1990	2000	Conditional on measures to achieve the target not affecting New Zealand's competitive advantage, being cost-effective, providing the greatest range of benefits whether or not climate change occurs, and providing a net benefit for New Zealand society.
Norway*	Target	CO_2	Stabilization	1989	2000	Preliminary.
Portugal**	EC Agreement	CO_2	(see footnote)	—	—	
Spain**	Target	CO_2	Limitation to 25% growth	1990	2000	Target approved by Parliament
Sweden*	EFTA Agreement	CO_2	(see footnote)			
Switzerland*	Target	CO_2	At least stabilization	1990	2000	Interim target.
Turkey	—	—	—	—	—	
United Kingdom**	Target	CO_2	Stabilization	1990	2000	Conditional on like action.
United States	Commitment to set of policies	All GHG	Stabilization	1990	2000	Stabilization achieved in part by CFC phase out
EC	Target	CO_2	Stabilization	1990	2000	Target is for Community as a whole

KEY

*=	EFTA Member
**=	EC Member
NMP=	Non-Montreal Protocol (refers to greenhouse gases other than those covered under the 1987 "Montreal Protocol on Substances that Deplete the Ozone Layer" and its subsequent Amendments i.e. greenhouse gases other than CFCs, HCFC, halons, carbon tetrachloride, and methyl chloroform).
GHG=	greenhouse gases
GWP=	global warming potential
Note:	EC Agreement means that country falls under EC-wide Target but has not yet developed its own target; EFTA agreement means that country falls under agreement between EFTA and the EC that EFTA members would together meet the EC target.

Source: IEA Secretariat and country submissions.

Table 2: **Carbon, CO$_2$, or Related Taxes in OECD Member Countries**

Country	Tax in Original Units for Main Products	Tax in $/TC1	Fuels Covered	Effective Date	Exceptions	Effects on Fuel Prices	Comments
Denmark	*Private* DKr 242/t of coal DKr 0.10/kWh DKr 320/t of fuel oil DKr 270/m³ of heating oil DKr 1.70/l of diesel oil *Industry* DKr 121/t of coal DKr 0.05/kWh DKr 160/t of fuel oil DKr 135/m³ of heating oil DKr 0.85/l of diesel oil	*Private* 15.8 *Industry* 7.9	*Private* Coal Oil but not gasoline Electricity *Industry* Coal Oil but not gasoline Electricity	*Private* 15/5/92 *Industry* 1/1/93	For energy-intensive industry, refunds up to 100% if reasonable conservation projects have been carried through	*Private* Coal up 5% Electricity up 3% *Industry* Coal up 16% Electricity up 12% Fuel oil up 18%	CO$_2$ taxes shown are part of an integrated CO$_2$ and energy tax system.
Finland	Mk26/TC	6.4	Fossil fuels	1/1/91	• Products used as raw materials in industrial production • Fuels in overseas planes and vessels	+ 1-2% for electricity, light fuel oil and natural gas + 5-8% for coal, gasoline and heavy fuel oil + 10% for diesel	Tax rate for motor fuels is larger than if it were proportional to carbon contact. Carbon tax was first instituted 1/1/90*l*. In 1991 all fuels taxes increased by 5% except those on motor fuel which increased 20%.

TC = Tons of carbon.
t = metric ton.

Sources: Country submissions.

1. Sweden based on first quarter 1991 exchange rates; Norway and Denmark based on last quarter 1991 exchange rates; Finland and Netherlands based on third quarter 1992 exchange rates.

Table 2: (continued): **Carbon, CO_2, or Related Taxes in OECD Member Countries**

Country	Tax in Original Units for Main Products	Tax in $/TC[1]	Fuels Covered	Effective Date	Exceptions	Effects on Fuel Prices	Comments
Netherlands	Gld 5.70/tonne CO_2 Gld 0.44/GJ	12.5 for CO_2 only	Fossil fuels, including industrial fuel gas	1992	None except non-energy uses and international sea/air traffic	Modest for transport; other wise 10-15% increase	Previous general environmental tax restructured to 50% CO_2, and 50% energy-based and raised in 1992
Norway	NKr 0.8/l of gasoline NKr 0.3/l of diesel and fuel oils NKr 0.8/m³ of natural gas NKr 0.3/k of coal[2]	196 (gasoline) 66 (diesel) 196 (natural gas) 47-70 (coal)	Oil products, natural gas and coal[2]	1/1/91 Revised 1/1/92 except coal 1/7/92	• Fuels in all air and sea transport • Coal used as input to industrial processes	+ 10-14% for gasoline, diesel and light fuel oil + 15% for heavy fuel oil	Diesel and fuel oil tax not increased in 1992
Sweden	SKr 250/t CO_2	166	Fossil fuels	1/1/91	Cap on total energy-intensive industrial CO_2 and energy taxes paid • Electricity sector • International sea and air traffic • Biofuels	With accompanying tax changes and simultaneous drop in crude oil prices, gasoline and diesel remained roughly the same	Major tax reform in January 1991. Energy, sulphur and nitrogen taxes are also in effect
	Residential SKr 320/t CO_2 Industry SKr 80/t CO_2	212 53	Fossil fuels Fossil fuels	1/1/93	Same as above plus ethanol	Depends on base market price in 1993. Assuming same prices as beginning 1992, prices plus taxes could rise 5 to 13% for residential. Industrial prices plus taxes could drop 25 to 40%	For 1993, energy tax cancelled for industry sector and for ethanol

TC = Tons of carbon.

t = metric ton.

Sources: Country submissions.

1. Sweden based on first quarter 1991 exchange rates; Norway and Denmark based on last quarter 1991 exchange rates; Finland and Netherlands based on third quarter 1992 exchange rates.
2. Coal covered by tax only after 1/7/92.

Table 3: **Climate Change: Key Energy and CO$_2$ Emissions Data[1] for OECD Countries (1990)**

	TPES[1] Mtoe	TFC[1] Mtoe	TPES/ GDP ratio[2]	TFC/GDP ratio[2]	Energy-related CO$_2$ emissions (ktons CO$_2$)	Energy-related CO$_2$ emissions per capita[3]	Energy-related CO$_2$ emissions per unit GDP[4]	% of energy-related OECD emissions of CO$_2$[6]	% of energy-related World CO$_2$ emissions[5]	CO$_2$ emissions from biomass (ktons)	CO$_2$ emissions from bunkers (ktons)
Australia	85.02	55.32	0.45	0.30	274 000	16.02	1.46	2.63	1.27	16 400	2 070
Austria	22.36	18.77	0.29	0.25	57 200	7.41	0.75	0.55	0.27	9 480	0
Belgium	51.81	38.39	0.55	0.41	124 000	12.40	1.32	1.19	0.57	949	13 300
Canada	202.43	151.32	0.50	0.38	435 000	16.35	1.08	4.18	2.02	33 500	2 010
Denmark	18.26	14.08	0.29	0.22	56 100	10.92	0.90	0.54	0.26	4 000	3 080
Finland	25.73	20.25	0.40	0.32	58 600	11.76	0.92	0.56	0.27	14 600	1 810
France	219.34	142.55	0.36	0.24	384 000	6.80	0.64	3.69	1.78	15 500	8 150
Germany	366.76	240.80	0.47	0.31	1 039 000	13.05	1.34	9.99	4.82	8 959	7 970
Greece	24.11	17.05	0.66	0.47	81 000	8.00	2.23	0.78	0.38	2 160	8 050
Iceland	1.43	1.13	0.44	0.34	2 450	9.57	0.75	0.02	0.01	0	0
Ireland	10.51	7.33	0.45	0.31	33 100	9.46	1.42	0.32	0.15	46	57
Italy	156.33	121.19	0.32	0.25	411 000	7.13	0.83	3.95	1.91	4 260	8 580
Japan	433.28	303.16	0.26	0.18	1 060 000	8.58	0.63	10.19	4.91	184	16 500
Luxembourg	3.54	3.37	0.83	0.79	10 300	27.10	2.42	0.10	0.05	97	0
Netherlands	77.12	63.26	0.54	0.44	183 000	12.22	1.27	1.76	0.85	792	35 000
New Zealand	13.55	9.71	0.59	0.42	26 800	9.45	1.39	0.26	0.12	1 420	1 020
Norway	21.07	17.51	0.34	0.28	31 900	6.33	0.43	0.31	0.15	3 740	1 450
Portugal	15.87	12.33	0.61	0.48	42 900	4.37	1.66	0.41	0.20	4 500	1 950
Spain	91.18	63.59	0.44	0.31	227 000	5.83	1.10	2.18	1.05	1 870	11 700
Sweden	42.83	28.74	0.38	0.26	55 300	6.47	0.50	0.53	0.26	21 000	2 130
Switzerland	24.19	19.46	0.23	0.18	44 400	6.53	0.42	0.43	0.21	3 310	58
Turkey	43.81	33.28	0.62	0.47	133 000	2.33	1.90	1.28	0.62	31 000	380
United Kingdom	212.15	149.48	0.40	0.28	589 000	10.26	1.11	5.66	2.73	1 530	8 000
United States	1 871.20	1 340.00	0.41	0.29	5 020 000	19.97	1.09	48.27	23.27	248 000	92 000
OECD Total	4 122.53	2 915.05	0.39	0.28	10 400 000	248.32	27.54	100.00	48.21	427 337	225 226
European Community	1 246.98	873.43	0.42	0.29	3 180 000	127.55	16.24	30.58	14.74	44 648	105 791

1. All data are IPCC adjusted: including bunkers and peat, excluding biomass, and treating sequested carbon as suggested in the IPCC/OECD methodology.
2. GDP in billion US$ 1985. TPES and TFC per GDP shows TPES and TFC per US$1 000 at 1985 prices.
3. t CO$_2$ per person.
4. t CO$_2$ per US$1 000 at 1985 prices.
5. World total CO$_2$ emissions = 21 570 432 ktons.
6. OECD total CO$_2$ emissions = 10 400 000 ktons.

Table 4: **Climate Change: Key Energy and CO_2 Emissions Data[1] Top Thirteen Energy-Related CO_2 Emitters Outside OECD (1990)**

	TPES[1] Mtoe	TFC[1] Mtoe	TPES/ GDP ratio[2]	TFC/GDP ratio[2]	Energy-related CO_2 emissions (ktons CO_2)	Energy-related CO_2 emissions per capita[3]	Energy-related CO_2 emissions per unit GDP[4]	% of energy-related OECD emissions of CO_2[8]	% of energy-related World CO_2 emissions[5]	CO_2 emissions from biomass (ktons)[6]	CO_2 emissions from bunkers (ktons)[7]
Brazil[9]	91.68	79.55	0.36	0.32	223 000	1.48	0.88	2.00	1.03	144 000	2 380
China	662.00	499.91	1.59	1.20	2 400 000	2.11	5.78	21.49	11.13	181 000	6 140
Czech & Slovak Federal Republic[10]	68.49	45.64	1.69	1.12	214 000	13.65	5.26	1.92	0.99	1 630	0
India	178.08	125.22	0.63	0.44	594 000	0.72	2.10	5.32	2.75	257 000	714
Kazakhstan[11]	69.42	N.A.	N.A.	N.A.	232 000	13.86	N.A.	2.08	1.08	N.A.	N.A.
Mexico	118.95	89.01	0.60	0.45	321 000	3.72	1.63	2.87	1.49	21 900	1 470
Poland	98.25	66.03	1.41	0.95	358 000	9.40	5.15	3.20	1.66	2 540	1 390
Republic of South Korea	93.99	72.10	0.62	0.48	246 000	5.74	1.63	2.20	1.14	2 600	4 860
Romania[11]	59.89	42.61	N.A.	N.A.	169 000	7.27	N.A.	1.51	0.78	3 700	0
Russia[11]	856.00	493.34	N.A.	N.A.	2 400 000	16.19	N.A.	21.49	11.13	N.A.	N.A.
Saudi Arabia	73.86	48.08	0.73	0.48	202 000	13.59	2.00	1.81	0.94	N.A.	8 840
South Africa	99.48	45.38	1.63	0.75	336 000	9.52	5.53	3.01	1.56	11 800	2 990
Ukraine[11]	223.83	N.A.	N.A.	N.A.	659 000	12.70	N.A.	5.90	3.06	N.A.	N.A.

1. All data are IPCC adjusted: including bunkers and peat, excluding biomass, and treating sequested carbon as suggested in the IPCC/OECD methodology, although subject to data availability constraints.
2. GDP in billion US$ 1985. TPES and TFC per GDP shows TPES and TFC per US$1 000 at 1985 prices.
3. t CO_2 per person.
4. t CO_2 per US$1 000 at 1985 prices.
5. World total CO_2 emissions = 21 570 432 ktons.
6. No figures available for Kazakhstan, Russia, Saudia Arabia and Ukraine.
7. No figures available for Kazakhstan, Russia, Saudia Arabia and Ukraine.
8. Non-OECD total CO_2 emissions = 11 170 432 ktons.
9. Special treatment for Brazil — TFC estimation for non-crude as it is predominately bio-ethanol.
10. Czechoslovakia as it was in 1990.
11. TFC figures not available for Kazakhstan and Ukraine. GDP figures not available for Kazakhstan, Romania, Russia and Ukraine.

Table 5: **World's Largest Energy-Related CO$_2$ Emitters, 1990**

	CO$_2$ (% World CO$_2$ Emissions)	CO$_2$ (ktons)
United States	23.27	5 020 000
China, Russia	11.13	2 400 000
Japan	4.91	1 060 000
Germany	4.82	1 039 000
Ukraine	3.06	659 000
India	2.75	594 000
Canada	2.02	435 000
Italy	1.91	411 000
France	1.78	384 000
Poland	1.66	358 000
South Africa	1.56	336 000
Mexico	1.49	321 000
Australia	1.27	274 000
Republic of South Korea	1.14	246 000
Kazakhstan	1.08	232 000
Spain	1.05	227 000
Brazil	1.03	223 000
CSFR	0.99	214 000
Saudi Arabia	0.94	202 000
Netherlands	0.85	183 000
Romania	0.78	169 000
Belgium	0.57	124 000
Greece	0.38	81 000
Austria, Finland	0.27	58 600
Denmark	0.26	56 100

Source: Adjusted data from IEA (1990).

Table 6: **Emissions of Carbon Dioxide from the ex-USSR Republics[1], 1990**
(thousand metric tons carbon dioxide)

Armenia	19 900
Azerbaijan	62 300
Belorussia	119 000
Estonia	11 000
Georgia	20 200
Kazakhstan	232 000
Kirghizia	18 600
Latvia	17 800
Lithuania	30 900
Moldavia	29 900
Russia	2 400 000
Tadzhikistan	11 400
Turkmenia	37 400
Ukraine	659 000
Uzbekistan	122 000
TOTAL	3 750 000

Source: Adjusted data from IEA document IEA/NMC(92)9.

1. Emissions are calculated from the apparent consumption of primary fuels (hard coal, brown coal, crude oil and natural gas) and petroleum products. Peat is not included, except for Russia.

Figure 1
Effects of Stated OECD Policies on World CO_2 Emission Forecasts[1]

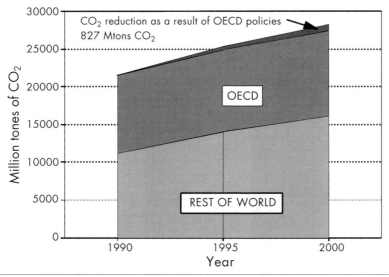

1. The underlying assumptions for the graph are as follows:
   - All EC countries are included under the EC target (stabilisation at 1990 levels by the year 2000).
   - For OECD countries other than EC members, individual country targets have been used in the calculations except for Sweden which has no stated target other than compliance with the EFTA target.
   - The US commitment has been calculated using USDOE figures for the emissions level in the year 2000, averaged between the Senate and House proposals for the National Energy Act, but adjusted for differences between the IEA and USDOE on emissions factors and emissions included or excluded in the methodology.
   - For Austria, Denmark, and Germany which have targets for the year 2005, it has been assumed, when converting these targets to the year 2000, that CO_2 emission reductions to meet the target started in 1990.

Effect of States Policies on Countries' CO_2 Emissions in Year 2000

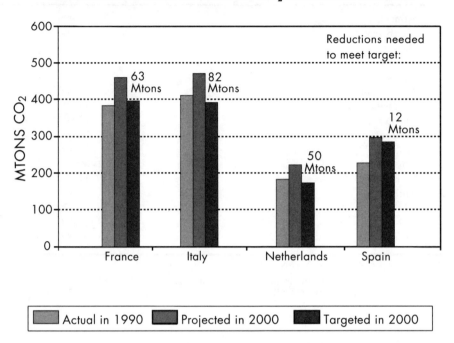

Figure 2b

Effect of States Policies on Countries' CO_2 Emissions in Year 2000

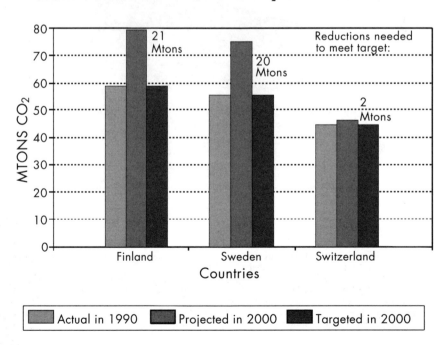

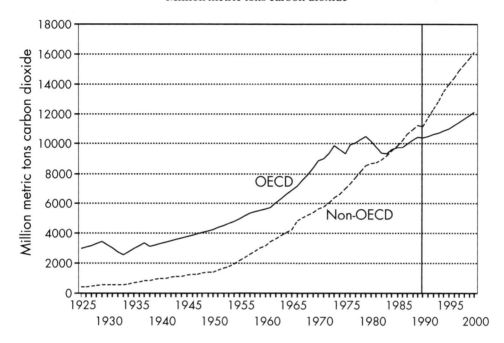

Figure 3
Total Carbon Emission Trends
Million metric tons carbon dioxide

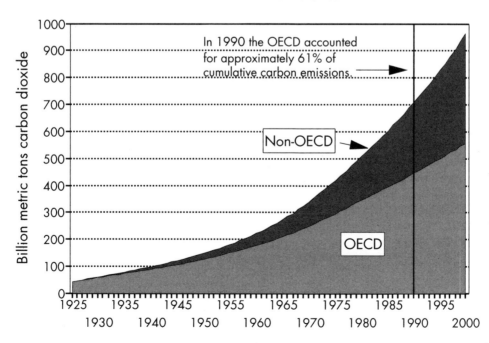

Figure 4
Cumulative Carbon Emissions
Billions metric tons carbon dioxide

SECTION 2

This section contains the country profiles for all OECD Member countries. Each profile contains four subsections: the official position of the country vis-a-vis energy-related climate change commitments, the factors contributing to the country's position, the relevant studies which have been completed by the country and key energy and CO_2 emissions data derived from IEA's energy balances using the IPCC/OECD methodology for CO_2 emissions inventories (see Section 1). As each profile has been approved by the Member country, it can be considered the authoritative description of each OECD Member country's national position.

In the Key Energy and Environmental Data sub-sections at the end of each country profile, it should be noted that the TPES reported and the derived CO_2 emissions are calculated using the IPCC/OECD methodology (see Section 1 for an explanation). Population estimates and the growth rates derived from them are taken from the United Nations *Monthly Bulletin of Statistics*, February 1992.

AUSTRALIA

1. OFFICIAL POSITION

Governments within Australia are committed to reducing Australia's greenhouse gas emission levels. A National Greenhouse Response Strategy is being developed by the Commonwealth, State and Territory governments in consultation with relevant industry, environment, and community groups. These commitments are affirmed in the Intergovernmental Agreement on the Environment of 1992.

As an early expression of these commitments, Commonwealth, State and Territory Governments adopted, at the end of 1990, an interim planning target to stabilise emissions of greenhouse gases by the year 2000 (based on 1988 levels) and to reduce those emissions by 20% by the year 2005. The interim target covers all greenhouse gases — notably, carbon dioxide, methane, and nitrous oxide — not controlled by the Montreal Protocol on substances that deplete the ozone layer. Australia has an existing policy of phasing out CFCs and halons by 1997.

Associated with this target is the policy caveat that no response measures will be adopted which have net adverse economic impacts nationally, or on Australia's trade competitiveness in the absence of similar action by major greenhouse-gas-producing countries. The target provides a focus for action in Australia to help forestall human-caused climate change, and a context within which planning can be undertaken; it is not legally binding.

New measures to increase domestic and industrial energy management information, introduce energy audits, and improve Commonwealth energy management practices, were announced in October 1990, and constituted the first substantial new action under the National Greenhouse Response Strategy. It is estimated that, if maintained, these measures can achieve savings in energy costs of A\$ 1.5 billion per year, with associated reductions in emissions of 14 million tonnes of carbon dioxide.

Subsequently, in September 1991, the Australian and New Zealand Minerals and Energy Council (ANZMEC), consisting of Australian Commonwealth, State and Territory, and New

Zealand Ministers with responsibility for energy policy, announced a suite of specific actions in the areas of energy labelling of domestic appliances, vehicle fuel efficiency, commercial building efficiency and increased industry involvement in energy efficiency.

Ecologically-sustainable development (ESD) remains a fundamental priority of the Commonwealth Government. As outlined in the Prime Minister's Statement on the environment of July 1989 and many times since, the Government aims to ensure that economic development is undertaken in an ecologically-sustainable fashion.

Nine Working Groups on ESD were commissioned in 1990 to develop strategies to increase the ecological sustainability of major sectors of the Australian economy. As part of their terms of reference, the ESD working groups were asked to identify the most cost-effective combination of measures available for reducing greenhouse gas emissions. These measures were consolidated and their potential collective impact analysed in the ESD *Greenhouse Report* (February 1992). Further, the Industry Commission was asked to report on the costs and benefits for Australian industry resulting from international consensus on a 20% reduction in greenhouse gas emissions. These reports have provided the basis for the actions proposed in the draft National Greenhouse Response Strategy (NGRS).

The ESD reports identified many actions to reduce greenhouse emissions which are cost-effective to the point that they return a net economic benefit, i.e. the savings pay for themselves. More generally, the ESD reports highlight *"no regrets"* actions, i.e. actions which offer environmental and/or economic benefits as well as greenhouse gas reductions.

The Heads of the Commonwealth, State and Territory Governments have announced their intention to give a joint response to the ESD reports in late 1992. This response will take the form of two parallel strategy papers: a National ESD Strategy and a National Greenhouse Response Strategy (NGRS). Drafts of these strategies were released as discussion papers at the end of June 1992 for public comment.

Draft National Greenhouse Response Strategy. The draft NGRS argues that the NGRS should be developed in a number of phases, with the first phase centred on no-regrets measures. It is recognised that early actions will also have to take into account budgetary, equity and burden-sharing considerations. Within the context of the Interim Planning Target, actions proposed in the draft NGRS are based on the recommendations of the ESD Working Groups. The Strategy recognises that a wide range of greenhouse-related actions and policies have already been adopted by the Commonwealth, States and Territories; many of the proposed new actions extend and expand on those in existing Commonwealth, State and Territory programmes, including the many programmes co-operatively organised through the Australian and New Zealand Minerals and Energy Council (ANZMEC) and the Australian and New Zealand Environment and Conservation Council. The Strategy provides for regular review through an intergovernmental (Commonwealth/State/Territory) body.

Australia is committed to a comprehensive approach to the greenhouse issue, i.e. addressing all sectors, all greenhouse gases, measures to limit emissions and enhance sinks, and adaptation and research. However, many of the first-phase actions proposed in the draft NGRS are in the energy sector, on the grounds of:

- significance of its contribution to greenhouse emissions;
- scope for no-regrets actions;
- measurability of emissions and reductions therein.

A major thrust of the first phase of the draft strategy is in measures to promote energy efficiency and conservation. Proposed new actions to improve the efficiency of energy use include:

- extension of the energy labelling scheme for domestic appliances;
- development of minimum energy standards for domestic appliances;
- development of a national House Energy Rating Scheme;
- development of energy performance standards for commercial buildings;
- development, where appropriate, of energy labelling and standards for industrial and commercial equipment;
- negotiation of a fuel economy programme for the car industry.

Other proposed first-phase actions include:

- greenhouse considerations to be taken into account in the National Grid Management Council;
- utilities to work towards effective integrated least-cost planning and demand side-management;
- strengthened RD&D on renewable energy systems;
- further studies on accounting for externalities in energy prices;
- a national programme to facilitate use of renewable energy stand-alone systems for remote areas.

No economic instruments such as a carbon tax are proposed as first-phase actions.

The draft NGRS also canvasses action on:

- transport (e.g. fuel efficiency targets);
- urban planning (actions already in the "better cities" programme, which have the effect of reducing the demand for transport and heating);
- improved management of extensive and intensive livestock (to reduce emissions of methane from this source);
- improved tillage practices (to maintain the effectiveness of soil as a greenhouse sink, and improve fertiliser efficiency and thus reduce emissions of N_2O);
- major tree-planting schemes;
- adaptation (mainly vulnerability and risk assessments);
- climate change research;
- economic research (especially assessments of proposed technical actions);
- community information and education.

International. Australia seeks to promote ecologically-sustainable development internationally. In the international negotiations leading up to signature of the Convention, Australia consistently sought an effective and equitable convention, with a comprehensive approach, covering all greenhouse gases and both sources and sinks.

At the UNCED, Australia signed both the Convention and the Biodiversity Convention.

2. FACTORS INFLUENCING DECISIONS

Australia is the world's leading exporter of coal, a major exporter of uranium and, since 1989, an exporter of natural gas. Indigenous production of oil currently provides the bulk of domestic requirements. Energy accounts for about 20% of Australia's export income. It is an important source of revenue for both the Commonwealth and State and Territorial governments and, consequently, is a significant element of national economic policy. Other sectors of the Australian economy, aspects of Australian life and the Australian environment are also very sensitive to potential climate change and/or to greenhouse response measures. For example, Australia is a significant exporter of alumina and aluminium; the Australian agricultural sector is very vulnerable to changes in rainfall and temperature; a significant part of the Australian population lives in coastal areas which would be threatened by a rise in the sea level; and much of Australia's unique natural flora and fauna would be endangered by climate change.

A critical aspect of Australian energy policy is the respective roles of the Commonwealth and State and Territorial governments, as well as the relationships among individual states and territories. The Australian Constitution gives the Commonwealth Government responsibility for taxation, trade, foreign investment, the development of offshore resources, the negotiation and implementation of international agreements and some other areas affecting energy policy. Constitutional authority over many energy-related areas — such as the regulation of electric and gas utilities, development and transport of onshore energy resources, and environmental protection — is assigned to the states and territories except in cases where a Commonwealth Government decision is required. However, the 1992 Intergovernmental Agreement on the Environment sets out a clearer division of responsibilities on environmental issues, including a framework for joint action on various issues, specifically including the development of a National Greenhouse Response Strategy.

3. RELEVANT STUDIES

R. L. Hawke, *Our country, Our future* (statement by the then Prime Minister, 1989).

Programs Implemented and Policies Adopted since 1988 that Contribute to Reducing Greenhouse Emissions in Australia (Australia and New Zealand Environment and Conservation Council), October 1991.

Ecologically Sustainable Development Working Groups, final reports on:

- Agriculture;
- Energy Use;
- Energy Production;
- Fisheries;
- Forest Use;
- Manufacturing;
- Mining;
- Tourism; and
- Transport.

(Nine volumes, AGPS, Canberra, December 1991; a separate volume of Executive Summaries was published at the same time).

Ecologically Sustainable Development Working Group Chairs, *Greenhouse Report,* Canberra, February 1992.

Industry Commission, *Costs and Benefits of Reducing Greenhouse Gas Emissions* (final report), Canberra, January 1992.

Australian Bureau of Agricultural and Resource Economics, Greenhouse: *Reducing Emissions from the Energy Sector,* Canberra, February 1992.

National Greenhouse Steering Committee, *Draft National Greenhouse Response Strategy,* Canberra, June 1992.

Australia

Key Energy and Environmental Data
(1990 data)

Adjusted TPES (Mtoe):	85.02
% Total OECD TPES:	2.06
% Total World TPES	1.08
Per capita TPES (toe per person):	4.89
Adjusted per capita TFC (toe per person):	3.24
TPES/GDP ratio (toe per US$1 00 1985):	0.45
TFC/GDP ratio (toe per US$1 000 1985):	0.30
Energy-related CO_2 emissions per capita (t CO_2 per person):	16.02
Energy-related CO_2 emissions per unit of GDP (t CO_2 per US$1 000 1985):	1.46
% Total energy-related OECD CO_2 emissions:	2.63
% Total energy-related World CO_2 emissions:	1.27
% Population growth (yearly average growth 1981-1990):	1.62

AUSTRIA

1. OFFICIAL POSITION

The Austrian Government supported the Action Plan of the Toronto Conference in 1988 which called for a reduction of CO_2 emissions by 20% of 1988 levels by the year 2005. The 1990 Energy Report, adopted by the government in June 1990, includes environmental compatibility as one of the major goals in Austria's energy policy and the Toronto targets for CO_2 reduction. Austria attended UNCED in Rio in June 1992 and signed the Convention.

The target for CO_2 is intended to be achieved through energy conservation and a shift from fossil fuels to renewable energy (hydropower, solar energy and biomass). A 15-20% energy saving potential has been identified and is currently being evaluated.

A conservation programme was launched in 1988. Financial support was cut back and more emphasis put on information and educational activities. The 1990 Energy Report states that the highest priority is given to energy conservation in Austria's energy policy. The conservation programme in the 1990 Energy Report puts more emphasis on market-oriented approaches and information. The main instruments to achieve the efficiency goals are research, energy auditing, creation of incentives and legal standards.

The 1992 Energy Report will be finalised in late autumn 1992.

Realised or planned measures to reduce emissions include:
- an energy tax is being discussed in order to further reinforce energy conservation and increase the use of renewable energy; the tax on oil products (except diesel) was increased on 1 January 1992;
- an optimisation of the electricity supply sector including new cost-oriented tariff structures;
- an exchange of research results among utilities in order to promote new technologies for energy conservation and renewable energy;
- a programme to promote the dissemination of new and environmentally-sound energy technologies;

- the reorganisation of the Energy Conservation Agency which acts as a clearinghouse for research and information activities;
- a pilot programme on photovoltaics and electric vehicles;
- a change in the tax system for motor vehicles; and
- a reduction of hydrocarbon losses and greenhouse gases from motor vehicles.

The Ministry for Economic Affairs is heading an inter-ministerial working group on energy policy responses to the problem of climate change.

Austria has signed the Sofia agreement and is thereby committed to stabilize the emissions of NO_x by 1994 and to reduce the emissions by 30% by 1998, compared with the 1987 level. In another international agreement Austria has agreed that by 1993 the national emissions of SO_x should be reduced to the 1980 level. During the last ten years, Austria has lowered its SO_2 emissions more than any other industrialised country. Austria signed the CFCs Protocol to the Vienna Convention. CFCs will be totally phased out by 1995.

2. FACTORS INFLUENCING DECISIONS

Hydropower is the main indigenous energy source, covering over 70% of electricity output. The 1991 referendum on the hydropower plant at Vienna ended up with a vote of 72% agreement. Oil covers 45% of the energy supply and gas 20%, both mainly imported. In 1978 a majority voted against the commissioning of a nuclear plant in a national referendum, which effectively halted the nuclear programme in Austria.

The federal nature of Austria's constitution means that both the federal (Bund) and regional governments (Länder) have responsibilities in energy and environmental policies. Examples of responsibilities of the regional governments are: building codes and air pollution from space heating.

3. RELEVANT STUDIES

Bundesministerium für wirtschaftliche Angelegenheiten (Ministry for Economic Affairs): *Energiebericht 1990 der Österreichischen Bundesregierung* (1990 Energy Report from the Austrian Federal Government). Vienna, 1990.

Österreichische Gesellschaft für Ökologie (Austrian Society for Ecology): *Energiesparpotentiale in Österreich* (Energy Conservation Potentials in Austria), Vienna 1991.

Energieverwertungsagentur (Energy Conservation Agency): *Dokumentation und Analyse der in vergleichbaren Industrieländern gewählten Strategie zur Erfüllung der CO₂-Reduktionsziele* (Documentation and Analysis of CO_2-Reduction Measures in Comparable Industrialised Countries), Vienna, 1992.

Österreichische CO_2-Kommission (Austrian CO_2-Commission): *Empfehlungen der Österreichischen CO_2-Kommission für ein Aktions-Program zur Erreichung des Toronto-Zieles und Jahresbericht 1991* (Recommendations of the Austrian CO_2 Commission for a Programme of Activities in Order to Achieve the Toronto-Recommendations and Annual Report 1991), Vienna-Laxenburg, 1992.

Under preparation:

Österreichisches Institut für Wirtschaftsforschung (Austrian Institute for Economic Research): *Investigation of Measures to Improve Energy Efficiency.*

Energieverwertungsagentur (Energy Conservation Agency): *Expertise on Policy Measures for Energy Conservation.*

Austria

Key Energy and Environmental Data
(1990 data)

Adjusted TPES (Mtoe):	22.36
% Total OECD TPES:	0.54
% Total World TPES:	0.28
Per capita TPES (toe per person):	2.90
Adjusted per capita TFC (toe per person):	2.43
TPES/GDP ratio (toe per US$1 000 1985):	0.29
TFC/GDP ratio (toe per US$1 000 1985):	0.25
Energy-related CO_2 emissions per capita (t CO_2 per person):	7.41
Energy-related CO_2 emissions per unit of GDP (t CO_2 per US$1 000 1985):	0.75
% Total energy-related OECD CO_2 emissions:	0.55
% Total energy-related World CO_2 emissions:	0.27
% Population growth (yearly average growth 1981-1990):	0.22

BELGIUM

1. OFFICIAL POSITION

The Government Declaration of May 1988 on the environment showed that Belgium is sensitive to protection of the environment. However, Belgium has no precise objectives or strategies to address the specific question of greenhouse gas emissions.

The general policy on environment would consist of pursuing energy efficiency improvements, an energy pricing policy taking into account the environmental aspects, continued R&D, continued nuclear energy and the development of combined heat and power.

Belgium considers that to deal with such an important and global problem as climate change a comprehensive approach in the EC framework and more generally in the international context (e.g. Toronto) is best.

The strategy to limit emissions of CO_2 and the possible introduction of an "ecotax" were discussed at the beginning of 1992 within the framework of a "Conseil National d'Avis sur le Climat, l'Environnement et le Développement" (National Council for Advice on Climate, Environment and Development) which participated in the preparation of the Rio de Janeiro Conference. This Conseil National is in favour of setting up a national target for the reduction of CO_2 emissions and targets for the reduction of other greenhouse gases. In addition, the Conseil National carried out an inventory of possible measures to comply with the reduction targets, the rational use of energy having a role to play in this strategy. Belgium attended UNCED in Rio and signed the Convention.

2. FACTORS INFLUENCING DECISIONS

The environment has become a priority in public opinion and in politics and each political party has developed an environmental plan.

The energy situation in Belgium is characterised by its important nuclear industry for electricity generation. Nuclear energy represented 60.2% of the fuel inputs in 1990 in electricity generation. The development of nuclear energy for electricity production has greatly contributed to the reduction of CO_2 emissions.

In the absence of new nuclear power stations which were to have been built in the framework of the Equipment Plan 1988-1998 for electricity generation, the Union des Exploitations Electriques et Gazières (Association of Electricity and Gas Producers, UEGB) points out that CO_2 emissions and the use of fossil fuels will increase and thus partly counteracts the limitation of CO_2 emissions which so far has been achieved. The building of combined cycle gas power stations can at best only slightly slow down this trend.

The Law of 8 August 1988 reorganises the responsibilities of the central government and the regions, effective 1 January 1989: its introduction has caused difficulties in establishing new environmental policies as the implementation of environmental regulations had been fully transferred to the regions. Procedures are made more cumbersome by the setting up of new structures in the regions. A co-ordination process in the field of environment is being considered to associate the Administration for Energy, the regions and the Secretariat of State for the Environment.

On 18 October 1991 representatives of the government and Belgium's two main electricity utilities — Electrabel and the Société Coopérative de Production d'Electricité (SPE) — signed a voluntary agreement on the reduction of SO_2 and NO_x emissions from conventional power stations. This agreement, covering 1993-2003, addresses four main areas: use of low-sulphur fuels, installation of sensing devices to monitor NO_x and SO_2 emissions from combustion chambers, completion of work on installing an FGD unit at the Alost plant in northern Belgium and engineering modifications to reduce NOx emissions from boilers.

Taking 1980 as the reference year, the agreement aims to achieve a 70% reduction in SO_2 emissions by 1993, a 75% reduction by 1998 and an 80% reduction by 2003. For NO_x emissions, the aim is a 30% reduction by 1993, a 40% reduction by 1998 and a 45% reduction by 2003. The two parties to the agreement are to set up a joint committee to monitor application of the agreement and, where necessary, to modify its terms according to changes in the economy from one year to another. The government has emphasised that the ground rules for this voluntary agreement, which is designed to provide a stable framework for the introduction of environmental protection measures by the electricity companies, are stricter than those set out in the EC Directive of 24 November 1988.

To meet the requirements for internal coherence and also to comply with international commitments, the various authorities responsible for environmental and energy policies are endeavouring to improve co-ordination with one another: the Ministry of Economic Affairs, Secretariat of State for Energy, External Relations, Public Health and the Secretariat of State for the Environment.

3. RELEVANT STUDIES

"Pollution atmosphérique dans le domaine de l'énergie" — Programme Energie — Convention No. 87-15, Francis Altdorfer. This study which is already available will be updated and data on CO_2 will be included.

A team of scientific researchers has been set up recently through a joint initiative from the Free University of Brussels, Liège University and the Belgian Space and Aeronautics Institute to identify the various atmospheric pollutants responsible for the climate changes and study their long-term developments, including a systematic survey of various monthly and seasonal variations in pollutant levels.

"Optimalisation de stratégies de réduction pour les émissions de gaz à effet de serre". This study is carried out in the framework of the national programme "Global change" of the Services de la Programmation de la Politique Scientifique. For this purpose, co-operation has been established between the Centrum voor Ekonomische Studiën (KUL) and the Vlaamse Instelling voor Technologisch Onderzoek (CEN/VITO). From 1991, these two bodies represent Belgium in ETSAP.

Belgium

Key Energy and Environment Data
(1990 data)

Adjusted TPES (Mtoe):	51.81
% Total OECD TPES:	1.26
% Total World TPES:	0.66
Per capita TPES (toe per person):	5.20
Adjusted per capita TFC (toe per person):	3.85
TPES/GDP ratio (toe per US$1 000 1985):	0.55
TFC/GDP ratio (toe per US$1 000 1985):	0.41
Energy-related CO_2 emissions per capita (t CO_2 per US$1 000 1985):	12.40
Energy-related CO_2 emissions per unit GDP (t CO_2 per US$1 000 1985):	1.32
% Total energy-related OECD CO_2 emissions:	1.19
% Total energy-related World CO_2 emissions:	0.57
% Population growth (yearly average growth 1981-1990):	-0.01

CANADA

1. OFFICIAL POSITION

In Canada's Green Plan, the Canadian Government committed itself to stabilize emissions of CO_2 and other greenhouse gases (not included under the Montreal Protocol) at the 1990 level by the year 2000. This is a national goal and does not directly pertain to specific regions or sectors. It is committed to phasing-out CFCs by 1995, methyl chloroform by 2000, and other ozone-depleting substances by the year 2005.

Canada is participating in international efforts to find a co-ordinated, equitable and effective set of actions involving developed and developing countries through the United Nations' negotiations on climate change. On 12 June 1992, Prime Minister Brian Mulroney signed the Convention on behalf of Canada at UNCED. Canada is committed to a prompt start for the Convention. Specifically, Canada has offered to sponsor:

- discussions on developing methodologies for implementing the comprehensive approach to greenhouse gas limitation; and
- an international meeting to discuss the funding mechanism contained in the Convention.

Canada has committed to ratify the Convention by the end of 1992. It has also committed to prepare, by June 1993, a national report outlining and assessing the impacts of policies and measures under way or planned along with an updated inventory and projection of greenhouse gas emissions. In order to support these undertakings, Canada will also promote a practical workplan for the Intergovernmental Panel on Climate Change to meet the scientific needs laid out in the Convention and will assist developing countries in meeting the scientific planning and report criteria through bilateral and multilateral support for country studies, transfer of technology and education, training and public awareness.

The Government of Canada has been working since 1990 with provincial and territorial governments to develop the National Action Strategy on Global Warming, a comprehensive framework for addressing the global warming issue within Canada. The federal and provincial governments are developing a three-part approach to climate change issues, namely to limit net emissions of greenhouse gases, to help Canadians anticipate and prepare for the potential effects of any warming that might occur, and to improve scientific understanding and increase predictive capability with respect to climate change. The federal, provincial and territorial governments are currently working to elaborate the National Action Strategy and ensure that environment, energy, economic and regional needs are balanced.

Canada's global warming strategy is national in nature, integrating the individual actions of Canadians and co-ordinating the activities of federal, provincial and territorial governments and all sectors of the Canadian economy. It is being managed through co-operative mechanisms and inter-government arrangements and agreements. A number of initiatives are already under way or planned in all jurisdictions across Canada. These are first step measures, many of which focus on improving energy efficiency across a broad spectrum of uses. The extent to which these actions will contribute to the national goal is being assessed. Further actions will be developed through consultative processes.

At this time, the Federal government has approved a new Efficiency and Alternative Energy (EAE) Program. The Program consists of a number of initiatives which are directed toward greater energy efficiency and the use of alternative energy in all end-use sectors — equipment, buildings, industry and transportation. They will employ a variety of policy instruments: information, suasion, research and development, and regulation. Those initiatives which have been announced are set out as below:

- National Energy Use Database. To improve the knowledge about the use of energy in Canada by organising and expanding a national database in this area;
- Energy Efficiency Standards for Equipment. To phase out inefficient equipment from the Canadian market by establishing national energy efficiency regulations for equipment imported into Canada or traded interprovincially;
- Energy Equipment Labelling. To require manufacturers to affix Energuide labels to equipment to indicate the associated energy consumption;
- Building Energy Code. To develop national model energy codes and specify minimum acceptable levels of thermal performance; and, promote their incorporation into provincial/municipal building codes;
- Building Information Transfer. To promote the adoption of energy-efficient technology by increasing public and industry awareness of energy efficiency opportunities and products available to the buildings sector;
- R-2000 Partnership. To promote energy-efficient houses and to train builders in the techniques required to design, construct and market these houses;
- Buildings R&D and Technology Transfer. To develop and commercialise energy-efficient and passive solar technologies in the residential and commercial sectors of the economy;
- Alternative Transportation Fuels — Technology and Market Development. To promote the development and market penetration of alternative fuel vehicles in Canada;

- Alternative Transportation Fuels — Research and Development. To promote research and development on alternative transportation fuels such as natural gas, propane, alcohol fuel and electric vehicles;
- Federal Buildings Initiative. To assist federal government departments to improve the energy efficiency of their facilities through a variety of techniques including savings-financing;
- Energy Innovators Ventures. To encourage major energy-using organisations to exploit energy efficiency and alternative energy opportunities by reaching agreements with them on performance and/or prescriptive initiatives they will undertake;
- Integrated Resource Planning. To improve the effectiveness of resource planning by electrical and natural gas utilities by encouraging the increased integration of demand-side management (DSM), non-utility generation, cogeneration, and district heating and cooling into the utility planning process;
- Industry Energy Efficiency. To enhance Canada's industrial competitiveness and help reduce emissions of CO_2 by increasing the efficiency of energy use in goods-producing industries; and
- Industry-Targeted Research and Development. To identify and foster opportunities for energy management research and development, field trials and technology transfer opportunities that could: improve energy efficiency; reduce environmental emissions; and improve industrial productivity.

Further initiatives will be announced in the areas of: information; transportation energy efficiency; and renewable energy. These new initiatives indicate that there are opportunities for a significant expansion of industries providing products, services and information in the area of energy efficiency and alternative energy (EAE). In addition, industries which undertake cost-effective investments in EAE will enhance their competitiveness.

Aside from energy efficiency improvements in the short term, alternative energies will be promoted, in particular less carbon-intensive energy sources, in order to contribute to the longer-term goal to reduce greenhouse gas emissions. The Canadian Government supports nuclear power as an option for electricity generation. Other measures include accelerated development and market penetration of alternative transportation fuels and the study of passive, active and photovoltaic solar energy options. Federal efforts will focus upon systems design and engineering as well as standard-setting and certification programmes, including enhanced R&D of alternative energy sources, cost-shared market assessments, and increased RD&D of advanced energy systems, e.g. combined cycle generation, CHP and district energy systems. These measures are to be supplemented by a major public information campaign and an annual report to Parliament on progress in meeting energy efficiency objectives. Other planned measures include a community tree-planting programme, a comprehensive emissions inventory and reporting system, as well as measures to limit greenhouse gas emissions from the agricultural sector and to reduce CFCs and ground-level ozone.

The government estimates that the outlined measures to limit carbon dioxide emissions are, of themselves, unlikely to realise the stabilization target. However, they will lay the foundation for achieving this objective, and will be supplemented by initiatives at provincial and territorial level. Possibilities include more aggressive demand-side management by electric utilities, changes in electricity pricing and regulatory structures, and systemic changes in urban centres, e.g. increased public transit and improved traffic flows. On the basis of continuing reassessment, additional measures will be considered. The use of economic instruments to achieve environmental objectives, including taxes and an emissions trading system, is under review. Finally, the government will significantly increase its commitment to scientific research on climate change, including more sophisticated climate modelling and monitoring. A discussion paper has been published by the Government of Canada to stimulate further debate and research into practical issues of design and implementation, and to identify points of focus for further work. Its purpose is also to serve as a basis for consultations with Canadians on the practical application of economic instruments to environmental problems in Canada.

Finally, the Government of Canada has established a Global Warming Science Program, which will focus the nation's expertise in climate change research and modelling. As part of that programme, Canada is developing a new and more sophisticated global circulation model to help formulate adaptive strategies and policy options. A climate Research Network will link the activities of various research centres around the world to each other. And, to ensure that research activities can build upon each other, a Climate Integration and Prediction Centre, with high-speed communications and computer resources, will act as the hub of the network.

2. FACTORS INFLUENCING DECISIONS

In the Canadian constitutional system, provinces have rights of resource ownership and legislative powers in wide areas of energy and environmental policy. The federal government has responsibility for overall economic policy and international and interprovincial trade. Federal, provincial and territorial governments co-operate closely in the matters of energy and environment, with regular meetings of the Federal-Provincial-Territorial Ministers of Energy, as well as the Canadian Council of Ministers of the Environment. The development of comprehensive regional and national action plans therefore involves a large degree of co-operative effort at all levels.

The energy sector plays an important role in the Canadian economy as a source of employment, national income, and export earnings. In 1990, the energy supply sector accounted for 6% of GDP and 14% of total investment. Energy-related activities provided about 3% of employment. Canada is rich in virtually all energy resources, including hydroelectricity, and is a net exporter of all the main energy commodities. It is the world's leading uranium producer and exporter. Currently, Canada sells about 42% of its energy production abroad. Energy exports generated about C$ 14 billion in revenue. Among the major energy users in the world, Canada has a relatively high ratio of energy use per unit of GDP. This energy-intensiveness is due to climatic conditions, geography-induced transport needs, a high standard of living, an industrial structure that reflects plentiful and competitively priced energy, and export of energy-intensive commodities.

Both energy supply and demand differ greatly among regions and provinces, as do levels of urban development and industrial activity. Consequently, carbon dioxide emissions vary both in quantity and source across the country. Ontario, the most populous and industrialised province, accounts for the largest proportion of Canadian emissions of carbon dioxide. Alberta, which produces large amounts of oil and gas and uses mainly coal to generate electricity, is the next largest contributor. Quebec is a relatively minor contributor of emissions in part because of its hydroelectricity generation. Regional variations in emissions are largely influenced by the way in which electricity is generated, by the proportion of electricity demand, and by the nature of the industrial structure. Fossil fuels are used extensively in electricity generation in the Atlantic region, Ontario, Saskatchewan, and Alberta. Ontario also makes extensive use of nuclear power. Quebec, Manitoba, and British Columbia rely heavily on hydro-electricity, and they tend to make more use of electricity, thus using less fossil fuel. One of the key end-use sectors is transportation which overall is a significant contributor to carbon dioxide emissions in all provinces.

3. RELEVANT STUDIES

Report on Reducing Greenhouse Gas Emissions. Federal/Provincial/Territorial Task Force on Energy and the Environment, August 1989.

Report of the Federal-Provincial-Territorial Task Force on Energy and the Environment, 2 April 1990, Kananaskis, Alberta.

The Green Plan — A National Challenge. A Framework for Discussion on the Environment. Environment Canada, March 1990.

A Report on the Green Plan Consultations. Environment Canada, August 1990.

Energy Use and Atmospheric Change — A Discussion Paper. EMR Canada, August 1990.

Climate Change Digest (Various publications by authority of the Minister of the Environment, Minister of Supply and Services Canada).

Canada's Green Plan — Canada's Green Plan for a Healthy Environment, December 1990.

National Strategy on Global Warming (Draft), released November 1990.

Economic Instruments for Environmental Protection: A Discussion Paper. Government of Canada, released in February 1992.

Canada and Global Warming: Meeting the Challenge. Government of Canada, released in June 1992.

Canada

Key Energy and Environmental Data
(1990 data)

Adjusted TPES (Mtoe):	202.43
% Total OECD TPES:	4.91
% Total World TPES:	2.58
Per capita TPES (toe per person):	7.60
Adjusted per capita TFC (toe per person):	5.68
TPES/GDP ratio (toe per US$1 000 1985):	0.50
TFC/GDP ratio (toe per US$1 000 1985):	0.38
Energy-related CO_2 emissions per capita (t CO_2 per person):	16.35
Energy-related CO_2 emissions per unit of GDP (t CO_2 per US$1 000 1985):	1.08
% Total energy-related OECD CO_2 emissions:	4.18
% Total energy-related World CO_2 emissions:	2.02
% Population growth (yearly average growth 1981-1990):	1.03

DENMARK

1. OFFICIAL POSITION

Denmark attended UNCED in Rio in June 1992 and signed the Convention there. Leading up to this, Denmark had been active in the intergovernmental negotiations on the Convention and has been a leader in taking national actions and policy initiatives related to climate change.

In response to the report of the World Commission on Environment and Development ("Brundtland Report"), the Danish government published in early 1990 two action plans, one on energy and the other on transport. The two action plans contain targets for the reduction of CO_2 emissions as well as other pollutants which were formally agreed to by the Danish Parliament in May 1990.

Corresponding to the Toronto targets, the agreed overall target for the energy sector, including the transport sector, is a reduction of CO_2 emissions by 20% in 2005 compared with the 1988 level, as well as further reductions in emissions of SO_2 and NO_x.

The energy action plan, which covers energy supply and demand in all sectors of the economy except transport, comprises a programme with initiatives to be implemented in the short term to reduce environmental impacts and to create a basis for further adaptation towards sustainable development in the energy sector. The government estimates that the following reductions can be achieved by the measures to be taken according to the energy action plan until 2005, as compared with 1988 levels:
- almost 15% in gross energy consumption;
- almost 30% of CO_2 emissions;
- around 60% in SO_2 emissions; and
- 50% of NO_x emissions.

The target of the transport action plan is:
- stabilization of energy consumption and CO_2 emissions in the transport sector before 2005, and a reduction of 25% until 2030; and

- reduction of NO_x and HC emissions of at least 40% before 2000, and further reductions thereafter.

The government estimates, however, that energy consumption and CO_2 emissions from the transport sector will be reduced by rather more than 5% by 2005 compared with the 1988 level. The government will follow up on the two action plans and monitor achievement of the targets, which are not mandatory. The energy action plan will be assessed in 1995 to decide whether the objectives or the means to achieve them need to be revised. The aim of both action plans is, first of all, to ensure substantial reductions in CO_2 as the most important greenhouse gas caused by energy activities in Denmark.

The energy action plan includes a comprehensive programme of action and is to be implemented by measures in four main areas: energy efficiency improvements and conservation in energy end-use, changes and improvements in efficiency in energy supply, increased utilisation of more environmentally benign energy sources, and R&D. Some important elements of the intended measures have already been approved by Parliament, e.g. a programme concerning the expansion of combined heat and power production, while other legislation is in preparation.

Energy taxes in Denmark are high, and in contrast to many other countries they cover both oil products, electricity, coal and (indirectly) natural gas. In December 1991 a proposal for CO_2 taxes was approved in Parliament. According to the Parliament decision, a CO_2 tax of DKr 100 per ton of CO_2, corresponding to about US\$55 per ton of carbon, will be introduced by 15 May 1992 on private energy consumption. The CO_2 tax will function in combination with the existing energy taxes, which will be somewhat reduced. For the industrial and commercial sector a CO_2 tax of DKr 50 per ton of CO_2 will be introduced by 1 January 1993. These sectors will as hitherto be exempted in general from ordinary energy taxes, and there will be certain possibilities for reductions in the CO_2 tax depending on the ratio of CO_2 tax relative to value-added in production. For industries in which the CO_2 tax will correspond to more than 3% of value added, there will be possibilities for a total tax reduction, if reasonable energy efficiency investments have been undertaken. According to the Parliament decision, the level of the CO_2 tax on industrial and commercial energy consumption will be increased gradually following the expected CO_2 tax decision in the EEC. Finally it has been decided in Parliament that a large part of the tax revenue will be used for certain specified energy efficiency subsidies, i.e. for promotion of DH and CHP and energy conservation in the commercial and industrial sector.

Efficiency gains are expected through revised and/or new standards for buildings, energy installations and appliances. Other measures in preparation in the area of energy efficiency include energy consultancy schemes, energy management in buildings, energy efficiency financing arrangements, energy labelling, programmes for public buildings, and co-operative arrangements with industry and utilities.

Initiatives planned to increase the efficiency of the energy supply sector include promotion of CHP in district heating, expansion of industrial cogeneration, connection of block heating centrals to the natural gas and district heat supply systems, use of natural gas in central power plants, and demonstration projects for coal gasification and fuel cells. Several measures are planned to increase use of renewable energies, in particular biomass, wind, and solar energy.

2. FACTORS INFLUENCING DECISIONS

Denmark has pursued an active energy policy since the 1970s in order to reduce its high reliance on oil. This was basically achieved through effective energy conservation programmes, development of indigenous energy resources (in particular oil, natural gas and renewable energies), a switch to coal as the primary fuel in electricity generation, a major expansion of CHP facilities and district heating systems, and a policy of introducing natural gas. Energy taxation policies traditionally have played an important role in Danish energy policy. The Danish Government also supports a substantial and comprehensive non-nuclear energy R&D programme.

Environmental consciousness in Denmark is high. Environmental aspects are an integral part of long-term energy policy, and the government has repeatedly given proof of its determination to make strong efforts to reduce the environmental impacts of the energy system. However, the energy action plan will necessitate quite substantial investments in all sectors, including industry. Not surprisingly, the industry association and some unions point out that risks are high for the competitiveness of the Danish industry as well as for employment.

3. RELEVANT STUDIES

Danish Ministry of Energy: *Energy 2000 — A Plan of Action for Sustainable Development*, April 1990.

Danish Ministry of Transport: *Action Plan on Transport*, May 1990.

Denmark

Key Energy and Environment Data
(1990 data)

Adjusted TPES (Mtoe):	18.26
% Total OECD TPES:	0.44
% Total World TPES:	0.23
Per capita TPES (toe per person):	3.55
Adjusted per capita TFC (toe per person):	2.74
TPES/GDP ratio (toe per US$1 000 1985):	0.29
TFC/GDP ratio (toe per US$1 000 1985):	0.22
Energy-related CO_2 emissions per capita (t CO_2 per person):	10.92
Energy-related CO_2 emissions per unit of GDP (t CO_2 per US$1 000 1985):	0.90
% Total energy-related OECD CO_2 emissions:	0.54
% Total energy-related World CO_2 emissions:	0.26
% Population growth (yearly average growth 1981-1990):	0.04

FINLAND

1. OFFICIAL POSITION

In 1990 the government prepared a report of current and planned national activities concerning sustainable development. This report was submitted to the Parliament in Autumn 1990. The report discussed climate change as the most important environmental issue for the future. This was the first time that Parliament had discussed the climate change issue.

Finland prepared a national report to the UNCED 1992. The report reflects the views of the Ministry for Foreign Affairs and the Ministry of the Environment on sustainable development and climate change.

In the beginning of November 1990 the government established a commission for elaborating strategies to control greenhouse gas emissions in Finland. The Commission reported its proposals at the beginning of June 1991. The work of the Commission was the first step in preparing a detailed national greenhouse strategy in Finland. As a next step a new commission (Carbon Dioxide Commission II) is working on a more detailed strategy by the end of 1993.

The Parliamentary Energy Policy Council published its proposal for an energy strategy of Finland in September 1991. In April 1992 the government prepared on that basis a White Paper on energy policy to the Parliament, named the Finnish Energy Strategy, where major policy questions are discussed including aspects relating to climate change. The Finnish Energy Strategy lists the following main objectives for energy policy: security of energy supply, efficient energy market and the limitation of the emissions (greenhouse gases included) of the energy sector. The report stresses the importance of stabilizing CO_2 emissions before the year 2000 and demand side management and energy conservation. The primary measures to achieve the objective of the strategy listed in the Finnish Energy Strategy include financial support to investments of new technologies, intensified research and development activities, regulations and norms as well as environmentally motivated taxes on energy.

A new approach to reduce the emissions in Finland is the use of environmental taxes. In 1990 the government proposed the imposition of a carbon dioxide tax which is estimated to raise the price of gasoline by 6%, the price of gas by 2% and the price of coal by 8%. In the budget for 1991 this tax was estimated to add (over 1990 levels) about 7% to the price of gasoline and about 5% to the price of other fossil fuels. In the 1993 budget these taxes are likely to be increased further. A task force has been set up to make an overall proposal on enhanced utilisation of environmental economic instruments by early 1993. On the other hand, taxes on environmental investments, such as sulphur and nitrogen oxide reduction investments of power stations, and clean vehicles have been reduced or eliminated.

In the view of the Finnish Government, climate change is an international problem that must be handled by international negotiations and agreements. Finland has endorsed the Noordwijk and Bergen declarations and the Ministerial Declaration of the Second World Climate Conference, thereby agreeing that a stabilization of greenhouse gas emissions is necessary as an initial step. The government signed the Convention during UNCED in Rio de Janeiro in June 1992.

Finland has signed the CFCs reduction Protocol to the Vienna Convention. The interim national goal for CFC reduction is 50% by the year 1992 compared to 1986 levels. The use of CFCs will be banned by the end of 1994. Halons will be allowed only for essential uses after 1991.

Finland has also signed agreements to limit emissions of SO_2 and NOx. Under the SO_2 agreement the national emissions of SO_2 are to be reduced by 30% by 1993 compared with 1980 levels. However by 1990 emissions of SO2 had already been reduced by about 60%. The national goal for SO_2 reduction (decided by the government January 1991) is 80% from 1980 levels by the year 2000. Finland has also signed the NO_x Protocol (Sofia 1988) and agreed to the goal for a NO_x reduction of 30% from 1980 levels by the year 1998.

2. FACTORS INFLUENCING DECISIONS

Finland's domestic fossil energy resources are limited. Approximately 30% of its energy needs are met by domestic supplies of hydropower, wood processing wastes, wood and peat. Almost 50% of the total energy consumed is derived from coal and oil. Nuclear energy at present provides some 15% of total final energy consumption.

Both per capita and per GDP energy consumption in Finland are considered to be relatively high. The climate and size of the country and the structure of business and industry can be considered the main factors influencing energy consumption. One fourth of the energy consumption is spent on heating of buildings, while industry accounts for nearly 50% of total final energy consumption. The products of forest based industry, the most energy-intensive industry, account for 40% of total exports.

More than 35% of all Finnish homes are connected to district heating networks. This figure will go up to 50% by the end of century. More than 60% of all district heat is produced in combined heat and power plants. The combined heat and power (CHP) potential and co-generation in industry are intensively utilised; the share of CHP of total electricity being the highest in the world. Because of this, the energy efficiency in urban areas is exceptionally high. On the other hand, the opportunities to decrease the energy demand and to decrease the CO_2 emissions by intensifying CHP or district heating are limited. Finland has already very low CO_2 emissions from electricity production.

Finland is very rich in forest and peatlands, both of which are sinks of carbon. The sustainable use of forest has been the basis of all forestry in Finland. The total amount of carbon in forest in Finland has increased during last 30-40 years. The area of forest has also increased slightly during last years. There is some evidence that it would still be possible to increase absorption of carbon in forests by means of forest management. The studies concerning the carbon balances in Finnish forests and peatlands are preliminary. More information is expected in two to three years, when the national research project on climate change (called SILMU) will produce assessment results.

Finland expects that it will experience large effects from climate change. The country is situated in the area where the temperatures are expected to rise significantly, especially in winter. The effects will most strongly be felt in the forestry sector and in the winter tourism sector. The risk of forest damage is expected to rise due to climate change and air pollution.

3. RELEVANT STUDIES

Most of the reports are available only in Finnish or Swedish.

Sustainable Development and Finland, Council of State Report to Parliament on Sustainable Development, Helsinki, 1990.

Report of the Carbon Dioxide Commission, Ministry of the Environment, Committee Report 1991:21, Helsinki, 1991.

The Finnish Energy Strategy, Council of State Report to Parliament on Energy Policy, Ministry of Trade and Industry, Reports C:31, Helsinki, 1992.

Finland, National Report to UNCED 1992, Publication by the Ministry for Foreign Affairs 13, Helsinki, 1991.

Greenhouse Gas Emissions Related to Energy Production and Consumption in Finland, Current Emissions and Some Future Technology Scenarios, Sture Boström, Rainer Backman, Mikko Hupa, Ministry of Trade and Industry, Research Reports D:197, Helsinki, 1991.

Greenhouse Gas Emissions in Finland 1988 and 1990, Energy, Industry and Transport Activities, Sture Boström, Rainer Backman, Mikko Hupa, Ministry of Trade and Industry, Ministry of the Environment, Turku June 1992 (to be published).

Economic Effects of Carbon Dioxide Emission Abatement, Ilmo Mäenpää, Hannu Tervo, Ministry of Trade and Industry, Reviews B:114, Helsinki, 1992.

Working Group Interim Report of Environmental Economic Incentives, 1990.

The Main Report of the Energy Conservation Project, Arto Lepistö, Ministry of Trade and Industry, Reviews B:100, Helsinki, 1991.

The Role of Peat Exploitation in Altering the Carbon Balance in Finland and Worldwide, Ministry of Trade and Industry, Research Reports D:183, Helsinki, 1990.

Finland

Key Energy and Environmental Data
(1990 data)

Adjusted TPES (Mtoe):	25.73
% Total OECD TPES:	0.62
% Total World TPES:	0.33
Per capita TPES (toe per person):	5.16
Adjusted per capita TFC (toe per person):	4.06
TPES/GDP ratio (toe per US$1 000 1985):	0.40
TFC/GDP ratio (toe per US$1 000 1985):	0.32
Energy-related CO_2 emissions per capita (t CO_2 per person):	11.76
Energy-related CO_2 emissions per unit of GDP (t CO_2 per US$1 000 1985):	0.92
% Total energy-related OECD CO_2 emissions:	0.56
% Total energy-related World CO_2 emissions:	0.27
% Population growth (yearly average growth 1981-1990):	0.44

FRANCE

1. OFFICIAL POSITION

Action Principles. France feels that in order to build up and maintain the general support essential for success, a certain number of principles must be observed:

- responsibility and preponderant role of the industrialised countries;
- the necessary participation of all countries from the very beginning;
- equitable differentiation of the commitments between countries or regions, taking into account levels of development, special circumstances or past efforts;
- the greatest possible harmonization of the measures adopted in order to avoid the distortions of competition; and
- consideration of the specific needs of the developing countries.

Establishment of Limitation Objectives. France believes that all gas emissions which contribute to the greenhouse effect should be limited but that carbon dioxide deserves special attention because:

- it is mainly responsible for the additional anthropogenic greenhouse effect;
- it remains for a very long time in the atmosphere once it has been emitted and this period becomes still longer as a result of the weakening of the ocean's role as a sink and as the warming effect increases;
- the reduction of CO_2 emissions as distinct from other gases requires important structural modifications in our economies, harmonized at the international level; and
- the quantities of CO_2 emissions from fossil fuels are the best known. Only the control of CO_2 emissions is practicable today, much in the same way as CFC emissions can be controlled.

Thus, even if the negotiations must take into account all the gases contributing to the greenhouse effect, it is inappropriate to seek to negotiate a global objective of limitation for all greenhouse gases, without running the risk of failing to adopt specific short-term commitments.

Due to an important nuclear power programme, and an important energy efficiency programme, France, which has already reduced its CO_2 emissions by 25% since 1980, has set itself a national stabilization target by the year 2000 at a level below 2 tons per inhabitant per year, provided that the major industrialised countries take a similar approach.

Adoption of Harmonized Instruments. It is commonly acknowledged that CO_2 emissions reduction requires "tough" policies at yet undetermined costs, particularly since the objectives established must be ambitious in order to meet the objective of stabilizing the warming effect. In this context, countries cannot undertake commitments unless they are assured that the actions decided will not create situations in which competition is distorted.

International harmonization of the economic and regulatory measures to be taken is essential. France welcomes the outcome of the Convention signed in Rio in June 1992, although this is only the first stage of a long process. Therein, the industrialised countries committed themselves to adopt national policies and measures to limit the effect of greenhouse gases by the end of the century, and recognised the necessity of returning, either individually or jointly, to the 1990 levels of carbon dioxide emissions and other greenhouse gases. Furthermore, these countries committed themselves to co-ordinate, if necessary, the appropriate economic and administrative tools in order to comply with these commitments.

France considers that, in the absence of an agreement between the industrialised countries on the principle of a tax based on carbon dioxide, priority should be given to harmonize progressively the taxation on fossil fuels through an adjustment on the countries with the highest level of taxation.

Moreover, the French experience shows that it is possible for industrialised countries to reduce appreciably their CO_2 emissions by a range of statutory actions and the choice of an appropriate energy policy.

As recommended in the final IPCC report, countries must initially consider the subsidies and tax incentives which favour the energy and greenhouse gas-producing sectors. France, for its part, believes that such an analysis would bring about the rapid abolition of fossil energy subsidies.

France further believes that the industrialised countries should institute within their national tax systems a graduated surtax on fossil energy at a uniform rate to cover the external costs of the greenhouse effect. This tax would be applied in the industrialised countries under conditions which would avoid both distortions in competition and the dislocation of industrial sites.

The implementation of actions designed to reduce CO_2 emissions would be the natural corollary of the signal sent by this pricing but it could not in itself build a genuine market for energy efficiency. Therefore, reaction to the price must be accompanied by the simultaneous implementation of the following actions:

- introduction of harmonized incentives or regulatory measures in the areas of industry, transport and construction; and
- emergence of a co-ordinated international effort for technological developments in energy management.

2. RELEVANT STUDIES

Rapport du Groupe interministériel sur l'effet de serre, November 1990.

Rapport de l'Académie des sciences, Autumn 1990.

France

Key Energy and Environmental Data
(1990 data)

Adjusted TPES (Mtoe):	219.34
% Total OECD TPES:	5.32
% Total World TPES:	2.79
Per capita TPES (toe per person):	3.89
Adjusted per capita TFC (toe per person):	2.53
TPES/GDP ratio (toe per US\$1 000 1985):	0.36
TFC/GDP ratio (toe per US\$1 000 1985):	0.24
Energy-related CO_2 emissions per capita (t CO_2 per person):	6.80
Energy-related CO_2 emissions per unit of GDP (t CO_2 per US\$1 000 1985):	0.64
% Total energy-related OECD CO_2 emissions:	3.69
% Total energy-related World CO_2 emissions:	1.78
% Population growth (yearly average growth 1981-1990):	0.46

GERMANY

1. OFFICIAL POSITION

In 1990, the Federal government decided to work towards a heavy reduction of CO_2 emissions in Germany as a precautionary measure. Because of the ecological and economical need for international action, it also decided to work towards international agreements for the protection of the climate, especially to reduce CO_2 emissions.

Being an important element of an overall strategy to deal with climate change, the federal government is developing an overall strategy to reduce CO_2 emissions. This strategy will consider international agreements and the effects on economic and social goals. An interministerial working group prepared a second report on possible measures to reduce CO_2 emissions. The Federal government accepted the report and asked the group to continue its work. The group will prepare further suggestions to reduce energy-related emissions of CO_2 — oriented at a 25-30% reduction by 2005 in Germany from 1987 levels — and to reduce other greenhouse gas emissions.

The concept will consider CO_2 reduction in the energy sector in general, in transportation, in the housing and residential sector, by new technologies and also in agriculture and forestry including sinks.

The government is preparing concrete measures:
- to use economic instruments with priority;
- to amend the energy law;
- to tighten standards for insulation and efficiency in heating systems;
- to improve education of professionals in construction related to energy saving;
- to increase energy efficiency in former East Germany; and
- to promote renewable energies.

CO_2 has the dominant attention, but CH_4, NO_x, N_2O and other greenhouse gases will be considered as well.

In addition to the above-mentioned concrete measures in preparation, Germany is considering the whole range of legal, economic and persuasive instruments. The main topics in further political discussion are, for example:

- measures to reduce energy consumption of motor vehicles;
- improvement and implementation of power plant technology towards better efficiency.

Germany attended UNCED in Rio and signed the Convention. The German Chancellor also invited the participants of UNCED to join the first follow-up conference of the climate convention (after ratification) to be held in Germany.

2. FACTORS INFLUENCING DECISIONS

German energy consumption relies heavily on the use of fossil fuels, especially on coal and lignite which are the indigenous energy sources in Germany (former FRG 30%; former GDR 70%). Although economic (hard coal) and environmental needs call for a smaller contribution of these fuels, social and regional considerations as well as security of supply aspects limit the timeframe and magnitude of the reduction potential. Improvement of energy efficiency in the former GDR has priority. At the same time countervailing effects such as increasing transport needs cannot be neglected.

Public awareness of environmental issues generally is very high in Germany, and it extends to the greenhouse gas discussion. It is difficult to foresee to what extent this awareness corresponds with the willingness to take or accept measures to protect the climate.

The timing of international agreements for climate protection will influence the timeframe and magnitude of measures in Germany.

The government, is pursuing an effective climate protection strategy, recognises that global problems require global solutions, that there is reciprocal dependence between the economy and the environment, and that unilateral action would adversely affect Germany's international competitiveness with little effect on global CO_2 emissions.

3. RELEVANT STUDIES

In 1987, the German Bundestag established a study commission on "Preventive Measures to Protect the Earth's Atmosphere" in order to deal with issues related to the growing threats to the earth's atmosphere. The Commission submitted three reports to the German Bundestag. The first offers not only a detailed account of current scientific knowledge about stratospheric ozone depletion and the anthropogenic greenhouse effect, but also recommendations on far-reaching measures to protect the earth's atmosphere (especially chlorofluorocarbons). In its second report the Commission studied the problems involved in protecting tropical forests.

The major area of the Commission's last report (published 5 October 1990, around 1 000 pages) is the avoidance and reduction of releases of radioactive trace gases due to energy use, and the possible content of an international convention for the protection of the earth's atmosphere. The Commission had embarked upon a comprehensive study programme which reveals the state of knowledge on these complex subjects.

The Commission asks — among others — for a reduction of CO_2 emissions in Germany by at least 30%, in the European Community by at least 20-25% and worldwide by at least 5% until 2005 from 1987 levels. This target for Germany should be reached:
- by a comprehensive adjustment of all energy-related laws and regulations;
- by a concept of energy taxes, incentives and measures in special sectors;
- by priority to energy saving and increased market penetration of renewable energies.

Germany

Key Energy and Environmental Data
(1990 data)

Adjusted TPES (Mtoe):	366.76
% Total OECD TPES:	8.90
% Total World TPES:	4.67
Per capita TPES (toe per person):	4.61
Adjusted per capita TFC (toe per person):	3.02
TPES/GDP ratio (toe per US$1 000 1985):	0.47
TFC/GDP ratio (toe per US$1 000 1985):	0.31
Energy-related CO_2 emissions per capita (t CO_2 per person):	13.05
Energy-related CO_2 emissions per unit of GDP (t CO_2 per US$1 000 1985):	1.34
% Total energy-related OECD CO_2 emissions:	9.99
% Total energy-related World CO_2 emissions:	4.82
% Population growth (yearly average growth 1981-1990):	0.21

GREECE

1. OFFICIAL POSITION

Greece has no stated position on greenhouse gas targets, nor any greenhouse gas stabilization or reduction programme. However, for some time, the Greek authorities have been working on the greenhouse effect issue. In view of the government, the problem of reducing greenhouse gas emissions would have to be dealt within the general context of the European Community. Greece attended UNCED in Rio and signed the Convention.

2. FACTORS INFLUENCING DECISIONS

Although Greece has no official position on greenhouse gas reduction, the protection of the environment has become a top priority in Greek politics. Protection of the environment is granted in the Greek constitution and a special court has recently been set up for this purpose.

In the main urban areas, Athens and Thessaloniki, where a large portion of the country's commercial and industrial activity is concentrated, the atmospheric pollution, particularly in summertime, reaches unbearable levels. The phenomenon (called Nephos in Greek) has recently taken a very large place in all the programmes of the political parties and in public opinion.

In February 1990, a comprehensive plan for the protection of the environment in the Athens area was finalised. With a budget of Dr 400 billion over a period of four years, it includes 50 actions to be implemented, including for example car parking and traffic policies, green belts and public transport systems. As the present government is aware that the situation requires an urgent solution, the emphasis of its actions has shifted to the transport sector which is now seen as the main contributor to air pollution because of: the increasing number of vehicles; the high average age of cars; poor maintenance of cars; inadequate transport networks, traffic control and parking spaces; and the inefficiency of public transportation.

It is estimated that 79% of pollutants in the Athens area are due to traffic pollution, mainly from old private cars — 2 million cars in this area have an average age of twelve years. In this context, in May 1989, a law introduced an incentive scheme to encourage the purchase of new cleaner cars with catalytic converters and consuming lead-free gasoline. It corresponds to an average reduction of 15% of the purchase tax on new cars. In addition, in February 1991, the new law 1921/91 was adopted by Parliament to accelerate the rate of renewal of the car fleet. According to this law, the following incentives are granted:

- reduction of the Special Consumption Tax by 60% for cars with a cylinder capacity up to 1 400 cm^3 and by 50% for cars with a cylinder capacity from 1 401 to 2 000 cm^3, but not more than Dr 2 000 000;

- abolition of the Additional Special Tax; and

- abolition of the road tax for five years for the new car.

The above incentives are granted in the case of purchase of a new clean passenger car or a light duty truck with a gross weight up to 2.5 tonnes and only on the condition that an old car is withdrawn from circulation. It is estimated that 400,000 polluting vehicles will be replaced by cleaner vehicles over a two-year period (1990-1992).

To accelerate the rate of renewal of the car fleet, consideration is being given to widening the scope of Law 1921/91: such incentives would also be provided for second-hand cars which are less than two years old equipped with catalytic converters.

Various specific measures to reduce pollution have been adopted — such as reduction of the sulphur content of heavy fuel oil and diesel oil, restriction of the lead content in gasoline — but they are judged absolutely ineffective by public opinion as atmospheric pollution increases continuously.

The most important indigenous production in Greece is lignite, mainly used for electricity generation. In 1989, lignite inputs represented 73.6% in electricity generation. Although there is no serious problem related to sulphur dioxide in northern Greece since the sulphur content of lignite is only 0.4-0.6% and, moreover, its high calcium content helps emissions to be kept to a low level, environmental considerations have resulted in a decision to import natural gas from the Soviet Union and Algeria. In addition, the strong interest shown by the Public Power Corporation in the development of renewable energy mainly in the islands — wind energy and geothermal — is also partly motivated by environmental considerations.

Greece

Key Energy and Environment Data
(1990 data)

Adjusted TPES (Mtoe):	24.11
% Total OECD TPES:	0.58
% Total World TPES:	0.31
Per capita TPES (toe per person):	2.38
Adjusted per capita TFC (toe per person):	1.68
TPES/GDP ratio (toe per US$1 000 1985):	0.66
TFC/GDP ratio (toe per US$1 000 1985):	0.47
Energy-related CO_2 emissions per capita (t CO_2 per person):	8.00
Energy-related CO_2 emissions per unit GDP (t CO_2 per US$1 000 1985):	2.23
% Total energy-related OECD CO_2 emissions:	0.78
% Total energy-related World CO_2 emissions:	0.38
% Population growth (yearly average growth 1981-1990):	0.45

ICELAND

1. OFFICIAL POSITION

Iceland has for a long time followed the policy of increasing the exploitation of relatively pollution-free hydroelectric and geothermal energy and minimising the consumption of fossil fuels. This policy is the basis of the Icelandic energy programme, both for the present and the foreseeable future.

At the Second World Climate Conference in Geneva in November 1990 the Government of Iceland agreed to restrict carbon dioxide emissions in the year 2000 to the emission levels of 1990. In relation to these aims, an Icelandic Carbon Dioxide Committee has been established, to examine how the aims can be fulfilled and to put forward proposals for emission controls. Iceland attended UNCED in Rio in June 1992 and signed the Convention. It hopes to ratify the convention in 1992.

Iceland is a party to the ECE Agreement on Long-Range Transboundary Air Pollution (Geneva, 1979), which took effect in 1983. Iceland confirmed its membership of the agreement in 1982. Because of the relatively low sulphur dioxide emissions in Iceland, the Icelandic Government has not considered it possible to sign annexes to the ECE agreement, regarding reduction of sulphur dioxide emissions below the present levels. However, Iceland has pledged to exercise every precaution concerning any new potential sources of sulphur dioxide pollution. Regarding emissions of VOC and NO_x, Iceland aims at confirming the ECE agreements in 1992.

According to a recently passed change on the regulation on pollution control, three-way catalysts must be installed in all new gasoline propelled cars sold after mid-year 1992. Thus, the NO_x emissions are expected to decrease at least 30% during the next ten years.

The government is currently considering the implementation of pollution taxes. A committee will be established soon to estimate the impact of pollution taxes and to work on proposals for controls through economic measures.

2. FACTORS INFLUENCING DECISIONS

Iceland is the smallest economy in the OECD and, by far, the smallest emitter of carbon dioxide; however, the per capita emission is relatively high. Hydropower provides 40% of the gross energy consumption, geothermal 32%, imported oil 25% and imported coal 3%. Thus, over two-thirds of the total primary energy is supplied by domestic, relatively pollution-free sources.

Traffic is responsible for two-thirds of the carbon dioxide emissions, primarily for fishing vessels, personal transportation and airplanes. Electricity is generated almost entirely from either hydropower (94%) or geothermal (5.8%). In order to reduce the strong economic dependence upon the fishing industry and to enhance stability of the Icelandic economy, Iceland is planning to increase the use of hydropower for electricity-intensive industry. Thus, non-energy-related emissions in Iceland could increase.

Presently there are relatively few industries in Iceland. The population is small and pollution has not created any major problems. However, the environmental consciousness in Iceland is growing very rapidly, and environmental aspects are of major pubic interest. Due to the fact that less than 30% of the primary energy comes from fossil fuel, and as personal and commercial transport consumes the bulk of the fuel, there are some technical limitations to reductions in carbon dioxide emissions in Iceland. The small size of the Icelandic market limits the options, e.g. regarding support for major R&D programmes to reduce emissions.

3. RELEVANT STUDIES

The Ministry of the Environment in Iceland, Icelandic Carbon Dioxide Committee, Draft Report 1991 (to be published soon; Icelandic).

Energy Prognosis Committee: Fuel-Prognosis 1988-2015 (Icelandic).

Iceland

Key Energy and Environmental Data
(1990 data)

TPES (Mtoe):	1.43
% Total OECD TPES:	0.03
% Total World TPES:	0.02
Per Capita TPES (toe per person):	5.60
Per Capita TFC (toe per person):	4.41
TPES/GDP ratio (toe per US$1 000 1985):	0.44
TFC/GDP ratio (toe per US$1 000 1985):	0.34
Energy-related CO_2 emissions per capita (t CO_2 per person):	9.57
Energy-related CO_2 emissions per unit GDP (t CO_2 per US$1 000 1985):	0.75
% Total energy-related OECD CO_2 emissions:	0.02
% Total energy-related World CO_2 emissions:	0.01
% Population growth (yearly average growth 1981-1990):	0.97

IRELAND

1. OFFICIAL POSITION

In January 1990 the government announced a new, comprehensive national Environmental Action Programme and a progress report on this programme was published in June 1991. As part of the programme, the government proposed the establishment of an Environmental Protection Agency, and the expansion of activities by other government bodies with responsibilities impinging on the environment. Legislation to establish the Environmental Protection Agency is being considered by Parliament at present.

The Environment Action Programme is based on the principle of sustainable development. Ireland participated actively in UNCED in Rio de Janeiro in June 1992 and signed the United Nations Convention.

Ireland is also actively involved in the development of the European Community's policy on climate change. The Department of the Environment is completing a national programme on climate change in association with the other government Departments involved (e.g. Energy, Finance, Agriculture and Transport). The programme includes measures on energy conservation as well as the development of carbon sinks.

The government believes that energy conservation is probably the most effective short-term contribution that can be made towards reducing CO_2 emissions and tackling global warming. To this end, as part of the Environment Action Programme, thermal insulation standards, and the range of buildings to which they apply, have been reviewed with a view to reducing further the amount of energy consumed on space heating. Additional resources were also made available in 1990 for energy conservation. These funds have gone towards a project to install energy management systems in two hospitals, to demonstrate the potential for savings in this area; a project examining the potential for extracting heat from ground water in the centre of Dublin; and a charitable group which draught-proofs the homes of the needy and elderly.

Ireland has also played a full role in action, under the Vienna Convention and Montreal Protocol, to eliminate emissions of CFCs and similar chemicals which account for up to 25% of the greenhouse effect. A major afforestation programme is planned for Ireland during the 1990s which will greatly increase national capacity to absorb CO_2 emissions.

The Department of the Environment is now working to develop, in association with other responsible Departments (e.g. Energy and Finance), further specific measures for meeting this challenge.

2. FACTORS INFLUENCING DECISIONS

Ireland is at the periphery of Europe and is poorly endowed with fossil energy resources. Per capita GDP in 1990 was 69% of the EC average. Its main indigenous energy sources are peat (turf) and natural gas. For oil it is dependent entirely on imports. The country's electrical and natural gas grids are isolated from the rest of Europe, though plans are under way to connect the gas grid via an underwater pipeline with the United Kingdom's natural gas system. Once this interconnection is complete, and with the likely prospect that new offshore deposits of natural gas will be found and developed, there is the possibility of expanding natural gas's share in all the consuming sectors. Over the longer term, there is a possibility that renewable energy sources, such as wind and wave energy, may be exploited on a large scale.

Ireland is remote from the major industrial areas of Europe and came late to industrialisation. It enjoys a relatively unspoilt environment. There is increasing awareness and public concern over environmental matters in parallel with the trend elsewhere in Europe. Indeed a number of warmer and drier summers brought attention to the sensitivity of the natural environment, particularly fresh-water fisheries, to slight changes in climate.

3. RELEVANT STUDIES

In December 1991, the Minister for the Environment published a collection of studies on the implications for Ireland of global climate change. The studies were based on the scientific analysis carried out by the United Nations Intergovernmental Panel on Climate Change and dealt with the following sectors — Agriculture, Forestry, the Green Mantle, Hydrology and Freshwater Resources, Coastal Areas and Fisheries. The studies necessarily represent only a first assessment of the significance for Ireland of change in climate patterns. Nevertheless, the studies point to a number of possible adverse implications, including deterioration of peatlands, modification of estuarine ecosystems, reduced water supplies in summer periods and increased incidence of flooding in coastal areas. The studies also indicate some favourable consequences, such as increased agricultural potential through wider production options and reduced costs. The studies have provided a useful input to the development of a national programme on climate change.

Ireland

Key Energy and Environmental Data
(1990 data)

Adjusted TPES (Mtoe):	10.51
% Total OECD TPES:	0.25
% Total World TPES:	0.13
Per capita TPES (toe per person):	3.00
Adjusted per capita TFC (toe per person):	2.09
TPES/GDP ratio (toe per US$ 1 000 1985):	0.45
TFC/GDP ratio (toe per US$ 1 000 1985):	0.31
Energy-related CO_2 emissions per capita (t CO_2 per person):	9.46
Energy-related CO_2 emissions per unit GDP (t CO_2 per US$ 1 000 1985):	1.42
% Total energy-related OECD CO_2 emissions:	0.32
% Total energy-related World CO_2 emissions:	0.15
% Population growth (yearly average 1981-1990):	0.19

ITALY

1. OFFICIAL POSITION

Italy signed the Convention in Rio during the UNCED. However the then Minister for Environment in his speech to the plenary session stated that 'without precise targets and timetables, the efficacy of the Convention is seriously impaired'. He called for a 'prompt start' to implementation of its clauses before ratification and an 'immediate start' of negotiations on protocols.

As for the other EC countries, Italy is engaged in carrying out the implementation of targets established by the joint Council of Ministers of Energy and Environment of 13 December 1991 who intend to stabilize CO_2 emissions at 1990 levels by the year 2000 for the European Community as a whole. Italy also intends to eliminate CFC production by 1997 and achieve a net forest growth by 1995.

The Italian Government has analysis under way regarding the possible effects of policy instruments such as incentives and taxation to reduce consumption and improve efficiency. Possible measures include taxes to reflect environmental impacts, use of more natural gas and other fuel substitution options and expanded exploitation of renewable sources of energy. The government believes that nuclear energy could play a role in the future with new and safer technologies, but its future use depends on specific conditions being met concerning containment, radiation releases and the solution of problems of radioactive waste and de-commissioning. (Italian voters approved three 1987 referenda questioning nuclear power as a significant contributor to the nation's energy mix.)

An updating of PEN 1988 is under way, which sets out a programme of activities aimed at improving energy efficiency and promoting energy conversion technologies that do not contribute to the build up of greenhouse gases. On 20 December 1990 the Italian Senate approved legislation enabling the government and the State energy bodies to carry out the Plan. The Plan provides for measures that are intended to affect consumer behaviour in the short term and those that will affect energy consumption in the medium term.

Short-term measures include a campaign to increase public awareness of the need to use energy rationally, and changes in pricing formulae in order to adjust cost increases as promptly as possible. It is also envisaged that part or all of any price decreases will be offset by higher taxes on energy. ENEL, the national electricity board, is expected to introduce time-of-use tariffs to the household sector. Rates and conditions for electricity purchased by ENEL from independent electricity generators operating CHP systems or using renewable energy sources have been made more favourable.

Measures intended to achieve energy savings and emission reductions in the medium term include financial incentives for energy-efficient investments, renewable energy sources, research on new nuclear-power technologies, and public transit. Besides investments, new rules are being drawn up to increase energy efficiency. These include new building codes, energy auditing services and installation of new light sources and controls in public buildings; periodic inspection of car efficiency, and enforcement of speed limits; and mandatory labelling of electricity consumption rates on household appliances.

2. FACTORS INFLUENCING POSITION

The rapid growth of the Italian economy, combined with roughly stable energy prices before August of this year, has led to continuing increases in energy requirements, especially in the electricity sector. The 5% growth in electricity consumption during 1988 forced a 35% rise in net electricity imports, despite a small increase in domestic production. The latest electricity demand figures show an additional 3.9% increase. Consumption of natural gas expanded 4.7% in 1988 and 8.4% in 1989.

Italy meets less than 20% of its total energy requirements, and less than 5% of its total oil requirements, from indigenous energy sources. Oil accounted for about 59% of the Italian energy supply mix in 1988, increasing to 62% in 1989. Energy-related and environmental issues have a high public profile, elevated by the recent oil market developments and by greater public awareness of various environmental impacts of energy-related activities.

3. RELEVANT STUDIES

Ministry of Industry and Trade, *National Energy Plan (PEN),* Rome, August 1988.

Ministry of Industry and Trade, *National Energy Plan (PEN),* Rome, November 1990.

Italy
Key Energy and Environmental Data
(1990 data)

Adjusted TPES (Mtoe):	156.33
% Total OECD TPES:	3.79
% Total World TPES:	1.99
Per capita TPES (toe per person):	2.71
Adjusted per capita TFC (toe per person):	2.10
TPES/GDP ratio (toe per US$1 000 1985):	0.32
TFC/GDP ratio (toe per US$1 000 1985):	0.25
Energy-related CO_2 emissions per capita (t CO_2 per person):	7.13
Energy-related CO_2 emissions per unit of GDP (t CO_2 per US$1 000 1985):	0.83
% Total energy-related OECD CO_2 emissions:	3.95
% Total energy-related World CO_2 emissions:	1.91
% Population growth (yearly average growth 1981-1990):	0.23

JAPAN

1. OFFICIAL POSITION

On 13 June 1992, Japan signed the Convention at UNCED in Rio de Janeiro. In his speech at the Earth Summit, Prime Minister Miyazawa reiterated Japan's intention "following its Action Program to Arrest Global Warming, to aim at stabilizing emissions of CO_2 by the year 2000 at about the 1990 level".

On 23 October 1990, the Council of Ministers decided on an "Action Programme to Arrest Global Warming" which covers the period from 1991 to 2010. Japan's basic policy position in dealing with global warming is based on three elements: the formation of an environmentally sound society, compatibility with a stable development of the economy, and international co-ordination.

In the Action Programme, the Japanese government established the following targets for the stabilization of Japan's CO_2 emissions, based on the common efforts of the major industrialised countries to limit carbon dioxide emissions:

- The emissions of CO_2 should be stabilized on a per capita basis in the year 2000 and beyond at about the same level as in 1990. To achieve this target, a wide range of measures under the Action Programme are to be steadily implemented, as they become feasible, through the utmost efforts by both government and private sectors.

- Efforts should also be made to stabilize the total amount of CO_2 emissions in the year 2000 and beyond at about the same level as in 1990. Among other measures, this is to be achieved through progress in the development of innovative technologies, including those related to solar, hydrogen and other new energies, as well as fixation of CO_2, at a pace and scale greater than that currently predicted.

Furthermore, the Action Programme stipulates that the emission of methane gas should not exceed the present level. To the extent possible, nitrous oxide and other greenhouse gases should not be increased.

With respect to carbon dioxide sinks, according to the Action Programme efforts should be made to work for the conservation and development of forests, greenery in urban areas, etc. in Japan, and also to take steps to conserve and expand forests on a global scale, among other measures.

The Action Programme outlines the necessary measures to achieve the above targets. To limit CO_2 emissions, measures are foreseen to achieve structures with reduced carbon dioxide emissions by reforming urban and regional structures, transport systems, the production structure, and the energy supply structure, as well as realising an appropriate life style. Measures to reduce methane emissions include the areas of waste management, agriculture, and energy production and use. Other measures are foreseen to reduce nitrous oxide emissions. Measures to enhance CO_2 sinks, e.g. forests and other greens, include adequate management of domestic forests and greens in urban areas, and the rational use of timber resources.

Additionally, scientific research and surveys as well as observation and monitoring are to be promoted. Further efforts concern the development and dissemination of technologies, e.g. for limiting emissions of greenhouse gases, for absorption, fixation, etc. of greenhouse gases, and for adaptation to global warming. Public awareness is to be promoted through, e.g., dissemination of information based on latest scientific knowledge on global warming, and the promotion of environmental education.

In the area of international co-operation on climate change, measures include comprehensive support and promotion of technology transfer, support to conservation and development of tropical forests and other carbon dioxide sinks, co-operation in R&D, promotion of international co-operation with private sectors, and international co-operative projects.

To implement these measures, the central government will provide support to local governments. The ministries and agencies responsible for the implementation of the Action Programme also are to support the efforts of the industrial and private sectors in the areas mentioned.

In addition, the Japanese Government is proposing to the international community "the New Earth 21", which is a dynamic and evolving strategy for international co-operation to restore the Earth over future decades through the reduction of greenhouse gases accumulated during the last two centuries since the Industrial Revolution.

Under this concept, the Japanese Government is already taking measures such as (1) to develop the innovative technologies such as CO_2 fixation and utilisation technology at the Research Institute of Innovative Technology for the Earth (RITE) and (2) to promote technology transfer to developing countries by training and dispatching experts through the International Center for Environmental Technology Transfer (ICETT) and other institutions, etc.

"The New Earth 21" consists of: (1) global promotion of energy efficiency and conservation; (2) massive introduction of clean energy sources; (3) development of innovative environment-oriented technologies; (4) enhancement of sinks; and (5) energy-related technologies that will carry us into the future. Its essence is (a) the promotion of world-wide diffusion of environment and energy technologies and (b) the development of innovative environment-oriented technologies.

2. FACTORS INFLUENCING DECISIONS

Japan imports over 80% of its primary energy requirements and in particular virtually all of its oil and natural gas. Although considerable progress has been made in reducing its oil dependence since 1973, mainly due to its successful energy efficiency and conservation as well as fuel switching programmes, it still has an oil share in TPES of 58% in FY 1990. Despite significant economic growth rates since the 1970s, its energy intensity (TPES/GDP and TFC/GDP ratios) is among the lowest in the OECD. Japan has one of the most comprehensive energy R&D programmes. The government's R&D budget reflects its strong intention to promote technological development.

The energy supply-demand outlook to 2010 announced in October 1990, reflects the government's determination to respond as effectively as possible to energy security requirements, to cope with increasing energy demand and to deal with global environmental problems. The outlook gives goals that should be attained through maximum efforts by both government and private sector, and it serves as a reference framework for more detailed supply plans such as those for electricity and oil and for other energy policy decisions. The outlook implies the following fundamental concepts. On the demand side, a vigorous effort at further improving energy efficiency and increasing conservation should be made, e.g. through significant improvements of the heating and cooling systems and automobile fuel efficiency, and further improvement of power generation efficiency. Concerning energy supply, the introduction of more non-fossil fuels is planned, particularly non-utilised and new energy sources, steady increase in nuclear power generation and introduction of cogeneration systems. The use of natural gas (LNG-based) will be increased for power generation and in the residential/commercial sector.

Public awareness of environmental problems is traditionally important in Japan, as the population has locally suffered from them earlier than in many other countries. Correspondingly, environmental restrictions in Japan have been among the tightest, with resulting encouraging improvements in environmental conditions. The government is making special efforts to raise awareness of global environmental problems to facilitate implementing measures.

3. RELEVANT STUDIES

MITI: *Long-Term Energy Supply-Demand Outlook,* June 1990.

Environment Agency: *Impacts and Response Strategies Concerning Climate Change;* Interim Report of Sub-Groups on Impacts and Response Strategies, The Advisory Committee on Climate Change, June 1989.

Japan

Key Energy and Environmental Data
(1990 data)

Adjusted TPES (Mtoe):	433.28
% Total OECD TPES:	10.51
% Total World TPES:	5.52
Per capita TPES (toe per person):	3.51
Adjusted per capita TFC (toe per person):	2.45
TPES/GDP ratio (toe per US$1 000 1985):	0.26
TFC/GDP ratio (toe per US$1 000 1985):	0.18
Energy-related CO_2 emissions per capita (t CO_2 per person):	8.58
Energy-related CO_2 emissions per unit of GDP (t CO_2 per US$1 000 1985):	0.63
% Total energy-related OECD CO_2 emissions:	10.19
% Total energy-related World CO_2 emissions:	4.91
% Population growth (yearly average growth 1981-1990):	0.56

LUXEMBOURG

1. OFFICIAL POSITION

The Luxembourg Government decided in November 1990 to stabilize the emissions of CO_2 at the levels of the base year 1990 by the year 2000 at the latest and to try to achieve a reduction of these emissions of 20% by the year 2005. Besides this and due to the geographic position of the country near large industrial areas and aware of the possible transfrontier pollution, the authorities are very much concerned about the problem of environmental protection and carefully follow the development of such policies in the context of the European Community.

The Ministry of Energy announced on 20 September 1990 an information/publicity campaign directed at energy consumers to strengthen their energy efficiency awareness which has been rather relaxed in the past context of low oil prices. One of the main arguments was that production, transport and use of energy have important impacts on the environment.

On 14 April 1992, Luxembourg adopted a law to come into force on 1 July 1992 that regulates the use of CFCs and other gases which attack the ozone layer. Luxembourg attended UNCED in Rio and signed the Convention.

2. FACTORS INFLUENCING DECISIONS

Energy conservation is considered by the Ministry of Energy not only as a way to reduce the country's dependence on imported energy but also to reduce efficiently emissions of a wide range of environmental pollutants. Although protection of the environment is not the only motivation for its recent action, the government decided this year to support investments in energy conservation and in the use of renewable sources of energy through financial measures. This concerns direct grants to homeowners for energy saving investments in existing buildings, direct grants for small-and medium-sized enterprises to help them finance energy audits and studies on possible energy efficiency measures, and direct grants for installations using renewable sources of energy or new technologies for energy conservation (solar energy, biomass, small hydropower, wind energy, heat pumps and CHP).

Luxembourg has already adopted different measures to limit the emissions of atmospheric pollutants, for example:

- fuel oil was completely phased out in the residential sector and replaced by gas oil with a maximum sulphur content of 0.2% by 1 January 1989;
- the sulphur content of heavy fuel oil used in combustion units rated at over 3 MW was limited to 1% from 1 July 1988 onwards and only gas oil with a sulphur content of 0.2% may be burned in combustion units rated at under 3 MW. For combustion units rated at 50 MW and above, the EEC Directive of 24 November 1988 is applicable;
- the iron and steel industry, ARBED, uses coking coal with a maximum sulphur content of 1%; and
- since 1 January 1990, new cars up to 2 000 cc fitted with a three-way catalytic converter qualify for a maximum subsidy of LF 20 000. The subsidy is limited to LF 10 000 for similar vehicles with oxidising converters. Older cars fitted with new catalytic converters qualify for the same grants.

Luxembourg

Key Energy and Environment Data
(1990 data)

Adjusted TPES (Mtoe):	3.54
% Total OECD TPES:	0.09
% Total World TPES:	0.05
Per capita TPES (toe per person):	9.30
Adjusted per capita TFC (toe per person):	8.85
TPES/GDP ratio (toe per US$1 000 1985):	0.83
TFC/GDP ratio (toe per US$1 000 1985):	0.79
Energy-related CO_2 emissions per capita (t CO_2 per person):	27.10
Energy-related CO_2 emissions per unit GDP (t CO_2 per US$1 000 1985):	2.42
% Total energy-related OECD CO_2 emissions:	0.10
% Total energy-related World CO_2 emissions:	0.05
% Population growth (yearly average growth 1981-1990):	0.30

NETHERLANDS

1. OFFICIAL POSITION

In May 1989, the Ministries with responsibilities for agriculture and fisheries, energy, environment and public housing, and transport and public works issued a joint White Paper on the environment, known as the National Environmental Policy Plan, or NEPP. Following a Parliamentary discussion on the paper, the government announced in November 1989 its decision to stabilize CO_2 emissions at the 1989/90 level by 1995 at the latest. In June 1990, a revised plan (the NEPP-Plus) was submitted to Parliament that calls for a 3% to 5% reduction on average 1989/90 levels by 2000.

The energy sector is expected to make the most important contribution (75%) to achieving these targets. Recycling and improved waste management is expected to account for 10% of the envisaged reduction, and the transport sector 15%. More than half of the necessary CO_2 reductions expected from the energy sector will be achieved through additional improvements in energy efficiency. This will require a rate of energy efficiency improvements of more than 2% per annum over the next decade — double the rate of improvement expected earlier for the 1990s. Details on how these goals are expected to be achieved were published in June 1990 in a White Paper on energy conservation.

The government will be relying on a mix of instruments to achieve its objectives. Publicity campaigns are already being expanded in an effort to increase public awareness and understanding of the global climate-change issue. The R&D budget for renewable energy has been raised from Gld 120 million to Gld 200 million a year. New building and appliance standards will be introduced, and existing ones will be tightened. Subsidies will be increased or reintroduced to promote solar energy, wind-generated power, combined heat and power, and other more efficient techniques, and to help defray the costs on retrofitting insulation and more efficient heating systems in existing buildings. Related subsidy budgets were raised from Gld 175 million to Gld 450 million. The government is also working out covenants with particular industries. The energy distribution utilities have proposed a major programme of investments to reduce pollution and improve energy efficiency within their areas of influence; the total cost of this programme will be Gld 250 million, which will be financed in part by a 1% to 2% increase in electricity and gas tariffs. Finally, a small tax on fuels (related to their CO_2 emissions) has been levied. This tax, which went into effect in February 1990, is intended more to raise revenue than to affect fundamentally consumer preferences. Since 1992 this tax, together with other environmental levies, has been reformed and considerably raised. Its level is comparable to the proposed EC CO_2/energy tax and its basis is 50% CO_2 and 50% energy content.

A Steering Group on Regulatory Energy Taxes has recently been set up to obtain an answer to the question: How far can regulatory energy taxes generate energy savings? Its report was finished in February 1992. Their study provides insight into the subsidiary effects of such an energy tax on such items as the distribution of incomes and purchasing power, the collective tax burden, employment, and so forth. The government has not taken a final position, waiting for the reports of its official advisory councils. Introduction is expected to be closely related and dependent on the negotiations and decisions concerning the EC proposal for the CO_2/energy tax.

While the government has taken these actions unilaterally, its position is that the problem of global climate change requires innovative international co-operation, with the main reductions in greenhouse gas emissions coming from the rich industrialised countries. The Netherlands is thus actively involved in the IPCC process and in multilateral programmes aimed at improving conditions in developing countries. The Netherlands attended UNCED at Rio and signed the Convention. In his speech to the Plenary Session, Mr. Alders, Minister for Housing, Physical Planning and Environment, announced the government's willingness to provide "new and additional financial resources up to a maximum of 0.1% GNP for the implementation of global environmental agreements, provided that other countries take a similar course in generating resources for such an earth increment".

On 4 September 1991 the Minister of Housing, Physical Planning and Environment has put forward to Parliament a White Paper on Climate Policy. The White Paper provides, in addition to an international vision, an integral picture of the global warming problem and the risks of climate change, as well as the Dutch effort at addressing the problem. A long-term environmental quality target has been set in addition to targets for the reductions of the emissions of greenhouse gases other than CO_2 and CFCs such as methane and N_2O. For N_2O the target is stabilisation of emissions in the year 2000 with respect to 1990 and for methane the target is a 10% reduction in 2000. Altogether the greenhouse gas reduction in 2000 will be -25% compared with 1990 levels. The White Paper also includes an analysis of the possibility of achieving a CO_2 reduction of 20% in the year 2005 with respect to 1988 (the Toronto target). The possibilities are studied in two ways: through an analysis of technological-oriented measures and through volume-oriented measures. The conclusions of this study are that international co-ordination on this target is necessary to reach this target for the Netherlands and that the bottlenecks are the social and practical aspects of implementation.

2. FACTORS INFLUENCING DECISIONS

The Netherlands is home to many energy-intensive industries and its competitive position is strongly influenced by the price of energy and particularly of electricity. The country is endowed with large deposits of natural gas, both onshore and offshore, and is a net exporter of this hydrocarbon. Natural gas has achieved a penetration ratio of over 80% of TFC in the household and commercial sectors, and almost 90% of TFC in the agricultural sector, where it is the main fuel used for space heating. Natural gas is also the mainstay of the electric power sector, accounting for half of the electricity generated in 1988. Efficient, natural-gas-fired combined heat and power (CHP) units, mainly integrated with industrial facilities, are expanding as a generating resource, and now account for over 15% of all the electricity generated in the country. Because of the high share of natural gas in the total energy mix, opportunities for reducing carbon emissions through fuel switching are limited.

Apart from fossil energy sources, the country's electric utilities generate each year around 0.8 Mtoe of electricity from nuclear power plants, and lesser, though growing, amounts from small hydro-electric plants and wind turbines. A *de facto* moratorium on the building of new nuclear power plants has been in effect for 15 years, and a resolution of the question is not likely until 1992 at the earliest. Earlier plans to raise the share of total electricity generated by coal-fired plants to 75% have been dropped for environmental reasons; a share closer to 50% is now envisaged.

The Netherlands' position at the heart of northern, industrialised Europe, its high population density, and its intensive agriculture, put considerable pressure on the country's environment; at the same time, public opinion polls show consistently the great importance that the population attaches to environmental matters. Concerns over the possibility of global climate change are rooted in part in the country's centuries-old battle against the sea. Government policies to reduce emissions of greenhouse gases enjoy widespread support. Most popular are those that encourage greater energy efficiency and the expansion of renewable energy. Opposition to nuclear power is still strong. Meanwhile, the government and the electric utilities are continuing to look into ways to make nuclear power plants even safer, and to improve techniques for safely storing radioactive waste.

3. RELEVANT STUDIES

Over ten reports have served as a basis for the government's proposals and subsequent Parliamentary debates. Some of the most relevant reports are listed below.

Minister for Public Housing, Spacial Planning and Environmental Protection; Minister for Economic Affairs; Minister for Agriculture, Natural Resources and Fisheries; and Minister for Transport and Waterworks, *National Milieubeleidsplan* [National Environmental Policy Plan], Report to the Second Chamber, Session 1988-1989, No. 21 137, The Hague, SDU Uitgeverij, May 1989.

McKinsey & Company, *Protecting the Global Environment: Funding Mechanisms,* report to the Ministerial Conference on Atmospheric Pollution & Climate Change, Noordwijk, Leidschendam, Ministry of Environment, May 1989.

Stichting Energieonderzoek Centrum Nederland, *Baseline and CO_2-response scenarios for the Netherlands,* Paper submitted to the IPCC, Petten, December 1989.

VEEN, VEGIN, and VESTIN, *Points of departure for the first environmental action plan of the energy distribution sector in the Netherlands: Integrated environmental policy plan for the Dutch energy distribution sector,* Arnhem, 26 April 1990.

Minister of Public Housing, Spacial Planning and Environmental Protection; Minister for Economic Affairs; Minister for Agriculture, Natural Resources and Fisheries; and Minister for Transport and Waterworks, *National Milieubeleidsplan Plus* [National Environmental Policy Plan—Plus], Report to the Second Chamber, Session 1989-1990, No. 21 137, The Hague, SDU Uitgeverij, June 1990.

Minister of Economic Affairs, *Memorandum on Energy Conservation,* Report to the Second Chamber, Session 1989-1990, No. 21 570, The Hague, SDU Uitgeverij, June 1990.

Minister of Housing, Physical Planning and Environment, *Memorandum on Climate Change,* Report to Parliament, September 1991.

Measures taken with the Netherlands' National Programme on Climate Change, CCD/Paper 3, April 1992.

Netherlands

Key Energy and Environment Data
(1990 data)

Adjusted TPES (Mtoe):	77.12
% Total OECD TPES:	1.87
% Total World TPES:	0.98
Per capita TPES (toe per person):	5.16
Adjusted per capita TFC (toe per person):	4.23
TPES/GDP ratio (toe per US$1 000 1985):	0.54
TFC/GDP ratio (toe per US$1 000 1985):	0.44
Energy-related CO_2 emissions per capita (t CO_2 per person):	12.22
Energy-related CO_2 emissions per unit GDP (t CO_2 per US$1 000 1985):	1.27
% Total energy-related OECD CO_2 emissions:	1.76
% Total energy-related World CO_2 emissions:	0.85
% Population growth (yearly average growth 1981-1990):	0.54

NEW ZEALAND

1. OFFICIAL POSITION

In July 1990, after receiving the recommendations of the New Zealand Climate Change Programme established in June 1988, the New Zealand Government agreed to aim for a 20% reduction of 1990 carbon dioxide emissions by 2005. The new government elected in October 1990 announced that this goal should be aimed for by the year 2000. In June 1992 New Zealand signed the Convention at UNCED. The New Zealand Climate Change Programme is comprehensive: it covers carbon dioxide, methane, nitrous oxide, both the sources and sinks of these gases, research, education, monitoring and international negotiations. It is also consultative: the development of policy is being, and will continue to be done in consultation with industry and other interested groups. In 1991, the government gave priority to climate change research by identifying climate change for a National Science Strategy (Climate Change).

Policy development has not reached the stage where specific instruments or measures are being recommended. This is seen as ongoing work and initial steps are only just being taken. However, it has been agreed that, in responding to the threat of climate change, New Zealand should begin by implementing measures which are estimated to be cost effective, to provide the greatest range of benefits whether climate change occurs or not, to have a net benefit for New Zealand society and to not reduce New Zealand's competitive advantage with its trading partners.

The Minister for the Environment released a Carbon Dioxide Action Programme in July 1992. This Programme is a first step towards reducing carbon dioxide emissions and focusses on measures that both reduce the net build-up of greenhouse gases in the atmosphere and/or increase carbon sinks, and have either a net economic benefit or low economic cost.

Although not a direct outcome of the Climate Change Programme, the government has directed the Building Industry Authority to develop residential and commercial building energy performance standards. The Ministry for the Environment is currently developing a comprehensive waste management strategy which will consider the reduction of greenhouse gases such as methane from landfill.

The Indigenous Forests Policy agreed to by the government in December 1990 has the primary objective to "maintain and enhance in perpetuity the existing areas of New Zealand's indigenous forests". New Zealand is also committed to the earliest practical phase-out of chlorofluorocarbons and was the second country to ratify the amendment to the Montreal Protocol which will see use of those substances cease by 2000.

2. FACTORS INFLUENCING DECISIONS

The decision to aim for a 20% reduction target for carbon dioxide emissions is placed in the context of international co-operation to achieve that goal. While New Zealand cannot on its own control the build-up of greenhouse gases in the atmosphere, the decision demonstrated New Zealand's willingness to take action, and is a signal to the international community of the importance attached by New Zealand to finding solutions to the climate change problem. New Zealand will continue to support international action on climate change through the Convention. Considerable concern is expressed in New Zealand at the possible effects climate change (especially rising sea levels) might have on Pacific islands with a largely coastal-based population.

3. RELEVANT STUDIES

New Zealand has contributed to the IPCC process and co-chaired the IPCC sub-group on Coastal Zone Management. The New Zealand Climate Committee of the Royal Society of New Zealand became the Facts Working Group of the New Zealand Climate Change Programme and prepared the *New Zealand Climate Report,* 1990 and an earlier abridged version *Climate Change in New Zealand,* 1988. The Impacts Working Group of the New Zealand Climate Change Programme prepared two reports, *Climate Change: a Review of Impacts on New Zealand* in April 1990 and the detailed background in *Climatic Change: Impacts on New Zealand* in May 1990. The Policy Working Group prepared *Responding to Climate Change: a Discussion of Options for New Zealand* in May 1990. The climate change policy response decisions were issued by the Prime Minister in August 1990 as *Climate Change: a Response Strategy.* In April 1991, the Ministry for the Environment completed a scoping paper on Developing a Strategy to Reduce CO_2 emissions, which sets out key issues and the associated work programme. The Ministry for the Environment prepared a report to government in May 1992 on measures to reduce or absorb carbon dioxide emissions that could be implemented at low economic costs or with net economic benefits, as the basis for the Carbon Dioxide Action Programme.

New Zealand

Key Energy and Environmental Data
(1990 data)

Adjusted TPES (Mtoe):	13.55
% Total OECD TPES:	0.33
% Total World TPES:	0.17
Per capita TPES (toe per person):	4.01
Adjusted per capita TFC (toe per person):	2.87
TPES/GDP ratio (toe per US$1 000 1985):	0.59
TFC/GDP ratio (toe per US$1 000 1985):	0.42
Energy-related CO_2 emissions per capita (t CO_2 per person):	9.45
Energy-related CO_2 emissions per unit of GDP (t CO_2 per US$1 000 1985):	1.39
% Total energy-related OECD CO_2 emissions:	0.26
% Total energy-related World CO_2 emissions:	0.12
% Population growth (yearly average growth 1981-1990):	0.82

NORWAY

1. OFFICIAL POSITION

The environmental policies and targets related to the energy sector were outlined in the White Paper on the Norwegian Government's follow-up to the World Commission Report on Development and Environment (Parliamentary Report no. 46). The Parliament gave its approval in April 1989.

At the same time, the Parliament stated that the emissions of CO_2 should be stabilized by 2000 at the 1989 level. This goal was preliminary and should be continuously analysed in the light of technological development, the outcome of further research and the result of international negotiations and agreements. Norway signed the Convention in Rio in June 1992.

Norway also signed the CFCs Protocol to the Vienna Convention. The national goal for CFC reduction is 50% by 1991 and at least 90% by 1995 (from 1986 levels). The use of halons is to be banned not later than from the mid-1990s, taking regard of international rules of safety and health.

Norway signed other international agreements to limit emissions of SO_2 and NO_x. By 1993, the national emissions of SO_2 should be reduced by 30% compared with the 1980 level. The objective of the government is to reduce the emissions by 50% by 1993, which was achieved in 1990. Its signature of the Sofia Agreement commits Norway to stabilize the emissions of NO_x by 1993 (from the 1987 level). In addition, Norway and twelve other Western European countries have signed a declaration aiming at 30% reduction of NO_x by 1998, compared with the 1986 level.

Since 1989, new gasoline-fuelled private cars have had to satisfy emission standards equivalent to the 1983 US standards, which require the use of catalytic converters. In October 1990, these standards were extended to vans and new diesel-fuelled cars.

In Norway, there has been an increased emphasis on economic instruments in the environmental policy. The objective is to introduce gradually a comprehensive pricing of emissions to air, which is as cost-effective as possible.

The government introduced a CO_2 tax on the domestic use of gasoline and mineral oils and the combustion of natural gas offshore in January 1991. Travel and transport by air, international and domestic sea transport are exempted from the CO_2 tax. The CO_2 tax was extended to include coal in July 1992. Coal, which is used as an input to industrial processes, was exempted.

Emissions of greenhouse gases is an international problem and must be handled by international co-operation and agreements based on global participation. The long-term objective should be to stabilize greenhouse gas concentrations at levels which secure a sustainable development of societies, minimise ecological damages and maintain climate conditions essential for the functioning of the biosphere. With a view to realising the long-term objectives, it is necessary to establish short- and medium-term global emission targets, and concrete commitments by participating countries in accordance with such quantitative targets.

The international efforts in order to limit the emissions of greenhouse gases should reflect the following operational principles:

- Precaution. In order to achieve sustainable development, climate policies should be based on the precautionary principle. Where there are threats of serious or irreversible damage, lack of full scientific certainty shall not be used as a reason for postponing measures to prevent unwarranted impacts;

- Cost-effective. Climate policies should be cost-effective to ensure global benefits at lowest possible costs;

- Equity. Policies to counter climate change shall be based on an equitable burden-sharing that recognises the "polluter pays" principle, the joint but differentiated responsibility of countries, differences in economic structures and resource bases and the circumstances of countries that will be abnormally affected by climate change and/or policies to counter climate change (e.g. exports of petroleum products); and

- Addition. Additional financial resources to cover incremental costs should be provided for developing countries and countries which will be abnormally affected by measures to counter climate change.

In order to achieve cost-effective solutions, climate policies should be comprehensive, include all relevant sources and sinks and comprise all economic sectors. Co-operation between countries should be stimulated and exchange of emission commitments between countries allowed. The marginal costs of reducing greenhouse gas emissions could be equalised through emission taxes and trading of emission rights.

As a first step, a mechanism for exchange of commitments could be established through a clearinghouse as a part of the Convention machinery and under the authority of the Parties.

The clearinghouse would appraise and select projects for reducing net emissions according to cost-effectiveness, and co-ordinate the funding of the projects from countries willing to undertake their commitments in co-operation with other countries. The net reduction in emissions resulting from any specific project should be credited to the contributing countries and deducted from their national commitments in accordance with agreement between the co-operating countries and subject to criteria approved by the Parties to the Convention. Such crediting should take place in the review process instituted as part of the Convention.

A clearinghouse as proposed could become an important supplement to other financial mechanisms and could become a very substantial new source of development funding through the transfer of financial and technological resources.

Norway considers the Convention to be a future-oriented convention, creating a good base for international co-operation on the principles mentioned above.

2. FACTORS INFLUENCING DECISIONS

As a major oil and gas exporting country, the Norwegian economy is closely related to the development in the international oil and gas markets. Much of Norway's projected growth in CO_2 emissions between now and the year 2000 comes from increasing production of oil and natural gas. Hydro power is the main indigenous energy source and covers 45% of the primary energy requirements and comprises 99% of the electricity generation. Oil covers 40% of the primary energy requirements, mainly in the transportation and manufacturing sector. A major part of the exports, excluding oil and gas, is energy-intensive products. The competitive position of the energy-intensive industries is strongly influenced by the price of energy, in particular electricity.

The production capacity of the operating hydro power stations and hydro power stations under construction is approximately 110 TWh in a normal year. The government has put forward a new plan to the Parliament regulating the further development of hydro power. According to the new plan, the production capacity would increase to 133 TWh/year if all projects were realised. This excludes water courses with a production potential of 34 TWh/year, which will be permanently protected for environmental reasons if the Parliament gives its approval to the plan.

In June 1990, the Parliament approved the new Energy Act. The Act is expected to lead to a significant reform of the electricity market, as a result of deregulation and increased competition in production and sale of electricity. These measures will lead to reduced costs, more rational investment decisions and a more flexible electricity trade. In addition, the reforms will contribute to a more efficient use of resources and improve the basis for evaluating investment costs against conservation costs.

State grants to energy conservation have increased over the last years. This covers conservation in housing, commercial buildings, manufacturing and service industries and the state's own buildings. It is estimated that the programme on energy conservation may gain more than 1 TWh/year including gains from extending and enhancing existing power stations.

3. RELEVANT STUDIES

Environment and Development. Programme for Norway's Follow-up of the Report of the World Commission on Environment and Development. Report to the Storting No. 46, 1988-89.

Energy Conservation and Energy Research and Development. Report to the Storting No. 61, 1988-89.

The Effect of different Measures to combat Emissions of Climate Gases on Energy Markets. Report by the Centre for Economic Analysis (ECON), 1990.

International Agreements on Reduction in Emissions of Carbon Dioxide. Report by the Centre of Applied Social Scientific Research, 1990.

Greenhouse Gas Inventory for Norway. Emission Amounts, Global Warming Potential and Emission Factors. Report by the State Pollution Control Authority, 1990.

Climate Change. The Effect of the Potential of Hydro Power and new Renewable Energy Resources. Report by the Norwegian Electricity Board, 1990.

The Effects of Climate Policies on Oil and Gas Markets. Report by the Center for Economic Analysis (ECON), 1990.

Energy, Environment and Development in China. Report by the Center of Economic Analysis (ECON), 1990 (in English).

Climate Policies in US, Soviet Union and EC. Three separate reports by the Fridtjof Nansen Institute, 1990.

Energy, Environment and Development in Brazil, Mexico and China. Three separate reports by the Fridtjof Nansen Institute, 1990 (in English).

A Study of India's Energy Situation and Climate Policy. Report by Nordic Center for Resource Studies, 1990 (English version available).

Effective verification of international agreements on Climate Change: Technically achievable but politically complicated? Report by the Fridtjof Nansen Institute, 1990.

International Agreements on Reductions in CO_2 Emissions. Report by R. Golombek, Center for Applied Research (SAF), 1990.

Energy Systems and Greenhouse Gases. Perspectives for Norway in the ETSAP-project. Report by Institute for Energy Technology, 1990.

Climate Change and Water Resources. Physical and socio-economical consequences of climate change on Norwegian water sources and water resources. Report by the Norwegian Water Resources and Energy Administration, 1990.

Development and refinement of transport models. Report by the Institute for Transport Economy, 1990.

Technological changes in energy efficiency and impacts on CO_2 emissions in the transport sector. Report by the Institute for Transport Economy, 1990.

Forests and Forest Production in Norway as a means to reduce atmospheric CO_2 concentration. Report by Department of Forestry, Agricultural University of Norway, January 1991.

Effects on the competitiveness of Norwegian Industries from an international climate agreement. Report by the Center for Applied Research (SAF), February 1991.

A national report on the use of environmental taxes. Preliminary report published in February 1991. Final report expected by the end of 1991.

The Greenhouse Effect, Impacts and Response Strategies. Report from the Norwegian Interministerial Climate Group, February 1991. (English version of Chapter 1, Summary, available; other main chapters will also be translated.)

The effects of a tax on CO_2 emissions on the Norwegian ferroalloys industry. Report by the Resource Strategies Inc., March 1991. In English.

Climate, Economy and Measures (KLOKT). Macroeconomic analysis. Report by the Central Bureau of Statistics, March 1991.

Bilateral Trade in Transferable CO_2 quotas. Report by the Center for Economic Analysis (ECON), April 1991.

Environment in the European Energy Charger. Report by the Center for Economic Analysis (ECON), June 1991 (in English).

"Towards a More Cost-Effective Environmental Policy in the 1990s". A report from an Inter-Ministerial Committee. NOU 1992:3.

Norway

Key Energy and Environmental Data
(1990 data)

Adjusted TPES (Mtoe):	21.07
% Total OECD TPES:	0.51
% Total World TPES:	0.31
Per capita TPES (toe per person):	4.97
Adjusted per capita TFC (toe per person):	4.13
PER/GDP ratio (toe per US$1 000 1985):	0.34
TFC/GDP ratio (toe per US$1 000 1985):	0.28
Energy-related CO_2 emissions per capita (t CO_2 per person):	6.33
Energy-related CO_2 emissions per unit of GDP (t CO_2 per US$1 000 1985):	0.43
% Total energy-related OECD CO_2 emissions:	0.31
% Total energy-related World CO_2 emissions:	0.15
% Population growth (yearly average growth 1981-1990):	0.38

PORTUGAL

1. OFFICIAL POSITION

There is no formal statement by the Government of Portugal on the consequences of energy use for the environment. More specifically, the government has not set any greenhouse gas emission targets. Nevertheless, as it had the presidency during the first six months of 1992, it was very active in the INC in negotiating and articulating the EC position. Portugal attended UNCED where it signed the Convention.

2. FACTORS INFLUENCING DECISIONS

Reflecting a growing awareness of environmental issues, a new Ministry for the Environment was created in 1990. Nevertheless, the government's efforts continue to focus on enhancing the performance of the economy. Protection of the environment and improving energy efficiency are two of the six general objectives of Portuguese energy policy, but these were not established with particular regard to the questions of CO_2 emissions and climate change.

As Portugal continues its efforts to diversify energy supplies and integrate into the European energy markets, its consumption of natural gas and coal will increase (at the expense of oil in certain sectors). As its economy expands, overall consumption of energy will increase (electricity, natural gas, transport fuels, and coal), so it is expected that CO_2 emissions will continue to grow at least until the year 2000. In the area of energy efficiency, the government will be launching a massive publicity campaign to increase the awareness of energy users of the need to improve energy efficiency. Improvements in energy efficiency will moderate the trend toward increasing CO_2 emissions.

3. RELEVANT STUDIES

Lei de Bases do Ambiente, in Diário da República, I Série, 7.4.90, pages 1 386-1 397.

Relatório do Estado do Ambiente e Ordenamento do Território, Ministry of Planning/ Ministry of Environment and Natural Resources, 1990.

Inventário das Emissôes de Poluentes Atmosféricos, Ministry of Environment and Natural Resources, 1990.

Cenários de Evoluçao da Procura de Energia. Ministry of Industry and Energy. Report prepared for the National Energy Plan, 1989.

Several studies on the formation and reduction of pollutants from combustion, including NO_x, N_2O, particulates. LNETI, Ministry of Industry and Energy.

Portugal

Key Energy and Environmental Data
(1990 data)

Adjusted TPES (Mtoe):	15.87
% Total OECD TPES:	0.38
% Total World TPES:	0.20
Per capita TPES (toe per person):	1.62
Adjusted per capita TFC (toe per person):	1.26
TPES/GDP ratio (toe per US$1 000 1985):	0.61
TFC/GDP ratio (toe per US$1 000 1985):	0.48
Energy-related CO_2 emissions per capita (t CO_2 per person):	4.37
Energy-related CO_2 emissions per unit of GDP (t CO_2 per US$1 000 1985):	1.66
% Total energy-related OECD CO_2 emissions:	0.41
% Total energy-related World CO_2 emissions:	0.20
% Population growth (yearly average growth 1980-1989):	0.76

SPAIN

1. OFFICIAL POSITION

The Spanish Government has set a target on CO_2 emissions from fossil fuels. A reduction of 20 points from the real trend is the objective of the National Energy Plan (PEN) that has been approved by government and ratified by the Parliament in April 1992. The target is to limit the emissions of CO_2 to a 25% increase between 1990 and 2000, instead of the rise of 45% which would be the real trend. This objective is inside the European Community's global policy towards stabilization of CO_2 emissions for the year 2000 at the 1990 level.

The continuing growth of the economy is likely to bring some degree of industry restructuring and a shift out of basic industry, with consequent environmental advantages. To meet objectives for atmospheric emissions in a cost-effective way, and consistent with an energy security goal, the following policies, considered in the PEN 1991, have been adopted: a Saving and Energy Efficiency Plan (PAEE) with specific programmes of energy saving, efficiency, cogeneration and renewable energies; an appropriate selection of new electric equipment; fuel-switching from high-sulphur local coal of low-calorific value to higher-quality imported coal, reducing consequently CO_2 and SO_2 emissions; and increasing the proportion of gas in the fuel mix.

Considering its future action on CO_2 emissions, the Spanish Government also refers to item 17 of the Noordwijk Declaration on Climate Change (6-7 November 1989) which states that industrialised countries with, as yet, relatively low energy requirements, which can reasonably be expected to grow in step with their development, may have targets that accommodate that development.

Spain attended UNCED in Rio and signed the Convention.

2. FACTORS INFLUENCING DECISIONS

When considering the level of CO_2 emissions, it must be borne in mind that the industrialisation of the country is relatively recent and that the potential of further economic growth remains important. The increase in electricity consumption and oil products for final energy consumption is expected to be above the average of the other EEC Member countries.

Following the oil crises in the 1970s, the Spanish energy policy adopted, as a priority, to increase coal use for both industry and electricity generation. Energy conservation and development of nuclear energy were two other priorities. However, the PEN of 1983 established a nuclear moratorium, which resulted in shelving plans to build five more nuclear reactors.

Another factor to be considered is the starting point of the CO_2 emission level per capita that is significantly smaller than the EC average.

Taking into account the above-mentioned factors and the EEC compromise of global stabilization of CO_2 emissions adopted in the Energy-Environment Council of 29 October 1990, the target set by the Spanish Government is to reduce 20 points from the real trend. Without the PEN 1991-2000 the emissions would rise 45% and with the measures of the PEN only 25%.

This target means an important effort and an important contribution to the EC target considering the starting point of Spanish emissions per capita.

Major RD&D efforts are being devoted to improving clean coal combustion technologies. These efforts were initiated to contribute to the reduction of SO_2 emissions, but they may also lead to lower CO_2 emissions per unit of energy produced. The main project is the Escatron pressurised fluidised bed combustion plant (PFBC) currently in operation.

3. RELEVANT STUDIES

Plan Energético Nacional 1991-2000 (National Energy Plan 1991-2000), 1991 (in Spanish).

Information for the Environment — Present Time and in the Future, Monography by the Directorate General of the Environment, Ministry of Public Works and Urbanisation (MOPU), 1989 (in Spanish).

Law on the Environment and its Guiding Principles, Directorate General of the Environment (MOPU), 1986 (in Spanish).

Guides for Methodologies for the Elaboration of Studies on the Environmental Impact, Directorate General of the Environment (MOPU), 1990 (in Spanish).

Spain

Key Energy and Environmental Data
(1990 data)

Adjusted TPES (Mtoe):	91.18
% Total OECD TPES:	2.21
% Total World TPES:	1.16
Per capita TPES (toe per person):	2.34
Adjusted per capita TFC (toe per person):	1.63
TPES/GDP ratio (toe per US$1 000 1985):	0.44
TFC/GDP ratio (toe per US$1 000 1985):	0.31
Energy-related CO_2 emissions per capita (t CO_2 per person):	5.83
Energy-related CO_2 emissions per unit of GDP (t CO_2 per US$1 000 1985):	1.10
% Total energy-related OECD CO_2 emissions:	2.18
% Total energy-related World CO_2 emissions:	1.05
% Population growth (yearly average growth 1981-1990):	0.36

SWEDEN

1. OFFICIAL POSITION

Sweden signed the Convention at UNCED held in Rio in June 1992. The Prime Minister urged that nations "start the process of presenting and discussing national plans" to meet the goals in the Convention "the sooner the better".

In January 1991, an agreement on energy policy was reached by the Social Democrats, the Liberal Party and the Centre Party. This agreement included a strategy for reducing climate change, which was presented to Parliament by the government in the Energy Policy Bill in February. The new four-party government installed in September 1991 has indicated support for the agreement even though the Conservatives and the Christian-Democrats were not party to the negotiations on the energy policy.

According to the agreement, Swedish efforts to limit climate change should be co-ordinated with those of other Western European countries and should contribute to action taken at an international level. The Swedish strategy to limit climate change must have a practical orientation and include all greenhouse gases and all economic sectors. Sweden must work actively to bring about emission reductions, even in sectors which are open to international trade and competition. Such far-reaching reductions require international co-operation.

The agreement emphasizes the efficient use of energy as an important instrument for limiting climate change and calls for an intensification of energy conservation efforts. Furthermore, environmentally-adapted energy production, with a relatively small impact on climate change, must be encouraged. Energy conservation measures and investments in environmentally-adapted energy production, combined with carbon taxes on fossil fuels, are to reduce carbon dioxide emissions from homes, services and district heating systems so that emissions in 2000 do not exceed today's levels.

The carbon tax was introduced on 1 January 1991 and applies to all fossil fuels except for fuels used for electricity production. The tax is reduced for energy-intensive industry and commercial greenhouses. Carbon dioxide emissions will be subject to a tax of SKr 0.25 per kg of carbon dioxide (US 4.1 cents/kg). This tax represents SKr 620/tonne of coal, SKr 535/m³ of natural gas, SKr 0.4/litre of LPG for cars, SKr 750/tonne of LPG for other uses and SKr 0.58/litre of gasoline. Emissions of carbon dioxide from domestic air traffic will be taxed at SKr 0.75/tonne of fuel.

In June 1992 the Parliament decided to change energy taxation. The new taxes will be in force on 1 January 1993. The carbon dioxide tax will be increased for the residential and the commercial sector to SKr 0.32 per kg of carbon dioxide and decreased for the industry sector to SKr 0.08 per kg of carbon dioxide. The energy tax will be maintained for the private and the commercial sector but cancelled for the industry sector. For energy-intensive industry the amount of CO_2 and energy taxes paid is capped for 1992 at 1.2% of total sales value of goods produced. This ceiling will be removed in stages in 1993 and 1994. Biofuels will continue to have neither energy nor carbon tax.

Furthermore an SO_2-emission tax was introduced on 1 January 1991 and a NO_x emission charge was introduced from January 1992. The NO_x charge is income-neutral as the revenue is redistributed to the energy producers, based on the production of electricity or heat.

Parliament has declared that the use of CFCs should be phased out completely by the year 1994. By the end of 1990, the use of CFCs in Sweden should be decreased by 50%, compared with the 1986 level.

2. FACTORS INFLUENCING DECISIONS

In the absence of national fossil fuel resources, Sweden has developed an energy system which relies heavily on electricity produced from nuclear and hydro energy. The Swedish economy is dependent on exports from energy-intensive industries and its competitive position is strongly influenced by the relative price of energy as compared to that of its competitors.

A ban on the further development of hydroelectricity in the four remaining large, unexploited rivers was adopted in 1987. Sweden has also set a target for a 80% reduction of SO_2 emissions in 2000 and of a reduction of NO_x emissions of 30% in 1995, compared with 1980 levels.

In February 1991, the Energy Policy Bill, based on the inter-party agreement described above, was presented to Parliament. According to the agreement, the time at which the nuclear phase-out can begin and the rate at which it will proceed will depend on the results of electricity conservation measures, on the supply of electricity from environmentally acceptable sources and on whether internationally competitive electricity prices can be maintained. The 1980 parliamentary decision to phase out nuclear power by 2010 has not been reconsidered.

The Energy Policy Bill sets forth an energy conservation programme as well as measures for promoting renewable energy production, including biomass and wind and solar energy. Taxation has been modified in order to make combined heat and power production more competitive.

3. RELEVANT STUDIES

Over twenty reports were completed in the course of the preparation of the Energy Policy Bill. Some of the most relevant reports published before 1992 are:

Environmental charges (Special Expert Committee).

Can industry meet environmental requirements? (National Industrial Board and National Board for Technical Development).

Proposals for planning the energy system to meet environmental requirements up to 2015 (National Environment Board and National Energy Board).

Consequences for electricity intensive industries of increasing electricity prices (Special Expert Committee).

More efficient use of electricity, electricity conservation and substitution (National Energy Board).

Economic analysis of domestic electric heating (National Energy Board and National Housing Board).

Need for improved information on electricity conservation and improved energy efficiency (National Energy Board).

Programme for air pollution control, including pollutants involved in acid rain and climate change (National Environment Board).

Reactor phase Out 1995/1996: Consequences and Social costs (National Energy Administration).

Growth and the Environment — a Study of Conflicting Objectives (Special Expert Committee).

Nuclear Phase-Out — Competence and Employment (Special Committee).

Electricity Market 1990 (National Energy Adminstration).

Some relevant reports published in 1992 are:

Energy in Sweden 1991 (Swedish National Board for Industrial and Technical Development).

Energy Report 1991 (R 1991;1, Swedish National Board for Industrial and Technical Development).

Climate Gases (Report 4011, National Environment Board).

Future Climate in the Nordic Region, survey and synthesis for the next century (SMHI Reports 64:1992).

Biomass Fuels — Effects on the Carbon Dioxide Budget (R 1992:10, Swedish University of Agricultural Sciences).

In addition, a number of studies were announced in the Energy Policy Bill. One of the most important studies is to be carried out by a Special Committee on Bioenergy, which was appointed in the spring of 1991. A study was in part carried out in November 1991 and the final report is expected in October 1992.

Sweden
Key Energy and Environmental Data
(1990 data)

Adjusted TPES (Mtoe):	42.83
% Total OECD TPES:	1.04
% Total World TPES:	0.55
Per capita TPES (toe per person):	5.00
Adjusted per capita TFC (toe per person):	3.36
TPES/GDP ratio (toe per US$1 000 1985):	0.38
TFC/GDP ratio (toe per US$1 000 1985):	0.26
Energy-related CO_2 emissions per capita (t CO_2 per person):	6.47
Energy-related CO_2 emissions per unit of GDP (t CO_2 per US$1 000 1985):	0.50
% Total energy-related OECD CO_2 emissions:	0.53
% Total energy-related World CO_2 emissions:	0.26
% Population growth (yearly average growth 1981-1990):	0.32

SWITZERLAND

1. OFFICIAL POSITION

The Swiss Government (Federal Council) has expressed its determination at the Second World Climate Conference in November 1990 in Geneva, to take measures for at least stabilizing the emissions of carbon dioxide until the year 2000 at the 1990 level. The country also signed the Convention on 12 June 1992 at the Earth Summit in Rio de Janeiro. Formal national targets and programmes already exist for such pollutants as NO_x, SO_2, HC and CFCs.

Presently implemented energy policy measures, which are also part of the "Clean-Air-Concept", are based on the "Energy Policy Programme of Cantons and Confederation" (1985), and on the Energy 2000 Action Programme launched in February 1991. They include, among others, cantonal building codes, federal standards for heating equipment and burner control, tax incentives, information activities, R&D and a programme for public buildings. An assessment of the progress achieved so far was published in September 1991.

In the field of transport, incentives for and investments in public transport, taxes on motor fuels and cars, speed limits and weight limits for trucks already contribute to reduce CO_2 emissions. In addition there are new projects for "piggy-back" transport of goods in international transit, "Rail and Bus 2000" and new railway tunnels through the Alps (AlpTransit).

After the constitutional amendment on energy was endorsed by public vote in September 1990, a Decree and an Ordinance on the Rational Use of Energy became effective in May 1991 and January 1992, respectively. These new legal bases enables the Federal government to considerably strengthen its policy.

The Federal Council can now implement, and has in part implemented, on a federal level, in particular, regulations on energy consumption of installations, equipment and motor vehicles, regulations on individual metering and invoicing of space heat and hot water consumption, regulations on the financial compensation of supplies for autoproducers to the public grid, and financial incentives for renewable energy and waste heat utilisation.

The objective of the Energy 2000 Action Programme is to increase efforts substantially, by broad participation of public and private actors, over the next ten years, in order to stabilize the consumption of fossil fuels and energy-related emissions of CO_2 between 1990 and 2000 at 1990 levels (thereafter consumption and emissions should be reduced). With regard to electricity demand, the programme aims at gradually reducing the growth rates and stabilizing demand as from the year 2000. The contribution of renewable energies is expected to increase (in particular by about 5% for hydropower). The performance of the existing nuclear capacities shall be expanded (by 10%) as far as safety requirements allow. Measures would include, on a federal level, the fast enactment of the decree on the Rational Use of Energy and, at a later stage, of an energy law, possibly a tax on CO_2, stronger financial incentives and additional and co-ordinated efforts of Cantons, communities and the private sector. The action programme constitutes an integral part of the environmental policies and programmes.

Regarding the possibility of a CO_2 tax, an interim report was prepared by the federal administration in 1990. The Federal Council will propose a Bill on CO_2 taxes. The timing of the implementation and the ultimate tax levels will depend on the outcome of the consultation procedure and on developments in the European Communities. For fiscal reasons an increase of gasoline taxes is being debated in the Federal Parliament.

2. FACTORS INFLUENCING DECISIONS

The new constitutional amendment on energy implies more competences of the federal state, in particular for regulations on energy efficiency and for financial support of renewable energies and waste heat utilisation. The new legislation will be essential for further increases in energy efficiency (and to a lesser extent for fossil fuel substitution) and thus for reduction of CO_2 emissions. The "Decree on the Rational Use of Energy" is planned to be replaced and supplemented and strengthened by the Energy Law in 1995.

Electricity supplies in Switzerland rely essentially on nuclear (41%) and hydro (57%). In the referendum in September 1990, a ten-year (and in fact even longer) moratorium regarding the licensing of new nuclear production facilities was approved. However, a gradual phase-out of nuclear energy was rejected. According to the federal government, the construction of fossil-fuelled power plants or additional long-term contracts for electricity imports should be avoided. Priority is therefore given to the efficient use of energy, including electricity. After the turn of the century, the question whether to renew, extend, or phase-out nuclear capacities will arise again.

Public opinion in Switzerland is very concerned about environmental protection, and the Swiss environmental legislation is one of the most advanced among IEA countries. Because of the country's complex political structure, energy and environment policies require close co-ordination among the federal, cantonal and community levels. The new constitutional amendment on energy and the Energy 2000 Action Programme should facilitate this co-ordination of energy policy. Switzerland has a system of direct democracy whereby citizens vote by referenda or initiatives on new laws or changes of the constitution. The implementation of climate change response strategies thus heavily depends on the public perception of the problem.

3. RELEVANT STUDIES

The federal government created in 1989 an interdepartmental working group within the federal administration, which has the task to co-ordinate ongoing activities and to advise the government regarding climate change issues. Studies on policies and perspectives of CO_2 emissions are being carried out within the Ministry of Transportation, Energy and Communication and are part of this work. An inventory of greenhouse gas emissions in Switzerland has been elaborated by the working group on the basis of available data on emissions of CO_2, CFCs, CH_4, N_2O and precursors of tropospheric ozone (NO_x, VOCs, CO).

Switzerland

Key Energy and Environment Data
(1990 data)

Adjusted TPES (Mtoe):	24.19
% Total OECD TPES:	0.59
% Total World TPES	0.31
Per capita TPES (toe per person):	3.56
Adjusted per capita TFC (toe per person):	2.86
TPES/GDP ratio (toe per US$1 000 1985):	0.23
TFC/GDP ratio (toe per US$1 000 1985):	0.18
Energy-related CO_2 emissions per capita (t CO_2 per person):	6.53
Energy-related CO_2 emissions per unit of GDP (t CO_2 per US$1 000 1985):	0.42
% Total energy-related OECD CO_2 emissions:	0.43
% Total energy-related World CO_2 emissions:	0.21
% Population growth (yearly average growth 1981-1990)	0.63

TURKEY

1. OFFICIAL POSITION

In Turkey, the sixth Five-Year Plan covering the period 1990 to 1994 takes into account environmental issues in every sector including energy. Among policies which have a positive impact on the environment are the following:

- economic assessment of environmental factors in energy fuel cycles, from production to consumption;
- more emphasis on R&D and transfer of technology to limit pollution from existing and new electric generating plants, e.g. R&D on fluidised bed combustion technologies;
- increased use of natural gas in electricity generation and in residential heating;
- support for R&D programmes on renewable energies to increase their use, e.g. increased exploitation of all hydro sources;
- improvement of the quality of oil products and measures to prevent pollution caused by their use; and
- increased emphasis on energy efficiency.

There is no production of CFCs in Turkey; the demand is met through imports.

There are three sets of measures that have been taken in Turkey to limit conventional air pollutants some of which are greenhouse gas emissions:

- an Environment Law (1983);
- an Instruction for Preserving the Air Quality; and
- an Instruction for Fuel Saving and Heating Insulation in existing buildings and for the Diminution of Air Pollution.

In this context, there are severe restrictions in the energy production area, especially for the coal-fired power stations. A decision that any new coal- or lignite-fired power station should be equipped with flue gas desulphurisation was confirmed in 1990. Such restrictions have been installed for conventional air pollutants some of which are greenhouse gas emissions.

Turkey attended UNCED in Rio in June 1992 but did not sign the Convention. The Minister of Environment stated to the Plenary session that "Turkey agrees with the target of reducing especially CO_2 and other greenhouse gas emissions which cause global warming and climate change" and also that "differences between countries must be taken into consideration when it comes to the sharing of responsibilities and sacrifices towards the solution of the global environmental problems". However, he also noted that there is a "necessity of determining certain criteria in the climate change convention regarding the level of development and emissions for the countries in accordance with scientific, objective and fair principles" [and] ... "population, surface area, wetlands and forests should be considered as the main parameters. Turkey believes that the Convention in its present form does not base on sufficiently explicit and objective criteria".

The new work programme of the General Directorate of Electrical Power Resources Survey and Development Administration (EIEI) which is in charge of analysing the efficient use of energy and research in new and renewable energy resources will be changed according to the results of the Rio Conference.

2. FACTORS INFLUENCING DECISIONS

Over the past few years, Turkey has experienced one of the highest rates of economic growth among IEA countries, averaging 6.1% per year from 1983 to 1988 and 1.0% in 1989. Nevertheless, Turkey has the lowest GDP per capita among IEA countries. Its population grew annually by 2.2% (slightly more than 1 million) on average between 1985 and 1990. An ambitious programme is being implemented both to modernise and privatise industry and to build the necessary infrastructure to boost energy supplies. Between 1975 and 1990, power generation capacity has been increased by a factor of 3.9, and a gas pipeline from the Bulgarian border to Istanbul and Ankara was built. Construction of an extension to an industrial area on the Black Sea has started in 1992. Turkey is developing a law to encourage geothermal energy utilisation.

The share of solid fuels in TPES in Turkey reached 45% in 1990, the highest among IEA countries (average about 25%). Domestic production accounted for over 71% of coal demand. Overall, indigenous energy production covered 45% of TPES in 1990. The government's energy policy goal is to ensure sufficient, reliable and economic energy supplies to support economic and social development. A major element is to increase domestic supplies, and, as far as possible, regional energy demand is to be met by regional production. With its large fossil fuel resources, Turkey plans to continue to construct new coal-fired power plants, and permission to construct and operate plants fuelled with imported coal according to the Build-Operate-Transfer (BOT) model was granted in 1986 but has not yet been implemented. For each plant the Ministry of Energy and Natural Resources must prepare a report evaluating the environmental effects. The share of hydro-electric capacity in electric generation is high (40%), and significant new hydro capacity is under construction. Taking advantage of small hydro-power resources, the goal is to reach a 50% share of electricity production by hydro.

Efforts to reduce air pollution in Ankara initially resulted in the gradual phasing out of the use of lignite in boilers which has been replaced by low sulphur imported steam coal. This move was initiated in 1986, and will be maintained until the use of natural gas will be available for the whole of the city, with the expansion of the distribution grid. At present the distribution network is under construction and the use of natural gas is available for nearly one-third of the city. Its use will be expanded on the basis of a master plan which is being pursued under the responsibility of Ankara Municipality. In addition some parts of the city of Istanbul will also begin to use natural gas for domestic and heating purposes soon.

The Prime Ministry Undersecretariat for Environment established in 1978, was converted into the Ministry of Environment in 1991. It has sole responsibility for planning, implementing and co-ordinating all environmental policy issues, including those related to energy in close co-operation with related ministries and government organisations.

3. RELEVANT STUDIES

A study project on "Assessment of the Principles of Legal, Administrative and Technical Measures for the Protection of the Ozone Layer" is in the 1990 programme of work of the Undersecretariat of Environment.

A national Climate Co-ordination Group has been set up in 1991 and a Sub-Group (Working Group on the Protection of the Atmosphere and Climate Change) prepared a national report on climate change and air pollution in May 1991. This report provides knowledge on greenhouse gases and air pollutants, but only CO_2 emissions in detail; and options for reducing or controlling them for Turkey.

Turkey

Key Energy and Environment Data
(1990 data)

Adjusted TPES (M(toe)):	43.81
% Total OECD TPES:	1.06
% Total World TPES:	0.56
Per capita TPES (toe per person):	0.77
Adjusted per capita TFC (toe per person):	0.58
TPES/GDP ratio (toe per US$1 000 1985):	0.62
TFC/GDP ratio (toe per US$1 000 1985):	0.47
Energy-related CO_2 emissions per capita (t CO_2 per person):	2.33
Energy-related CO_2 emissions per unit of GDP (t CO_2 per US$1 000 1985):	1.90
% Total energy-related OECD CO_2 emissions:	1.28
% Total energy-related World CO_2 emissions:	0.62
% Population growth (yearly average growth 1981-1990):	2.58

UNITED KINGDOM

1. OFFICIAL POSITION

The United Kingdom signed the Convention at UNCED. The United Kingdom fully supports the Convention as a significant first step in tackling climate change. At UNCED, the Prime Minister announced a new global Technology Partnership Initiative to promote technology co-operation in achieving sustainable development. It will be launched by a Technology Partnership Conference in the first half of 1993. The United Kingdom is pressing for the resumption of the work of the Intergovernmental Negotiating Committee to elaborate the Convention. The United Kingdom, along with other members of the G7 at the Munich summit, urged other countries to join them in seeking to ratify the Convention, and in publishing national plans for implementing it, by the end of 1993.

In April 1992 the government announced that the United Kingdom was prepared to bring forward its previous target and, provided other countries took similar action, to set itself the demanding target of returning CO_2 emissions to 1990 levels by 2000.

The United Kingdom's general strategy for limiting its emissions of greenhouse gases is set out in the Environment White Paper, "This Common Inheritance", published in September 1990. This covered areas such as energy efficiency, renewable energy, transport and nuclear power. In May 1992 the United Kingdom published a report to the European Commission on its national programme for limiting CO_2 emissions.

Energy efficiency measures and other measures which are worthwhile in their own right remain the priority. A major new initiative was announced in May 1992 for a partnership between the government, the energy utilities and OFGAS (the regulatory authority for the gas industry) to establish an Energy Saving Trust. The Trust will develop and propose programmes to promote the efficient use of energy by householders. OFGAS will allow the cost of approved schemes financed by British Gas to be recovered from tariff customers as a whole through an "E" factor. This and other initiatives to promote energy efficiency are expected to result in lower CO_2 emissions, amounting to some 3-5 million tonnes carbon by the year 2000.

The government remains of the view, however, that higher energy prices are likely to be needed in the longer term to encourage further CO_2 savings. The European Commission has now put a proposal to the Council for a carbon/energy tax. The United Kingdom will be taking forward discussion of this proposal during its Presidency of the European Community during the second half of 1992.

2. FACTORS INFLUENCING DECISIONS

The starting point for the UK Government's policies and proposals, as set out in the White Paper, is responsibility to future generations to preserve and enhance the environment of the country and the planet.

The fulfilment of this responsibility is based on a number of supporting principles, such as:

- policies must be based on and use the best scientific evidence and economic analysis available;

- where there are significant risks to the environment, the government be prepared to take precautionary action to limit the use or spread of potentially dangerous pollutants, even where scientific evidence is not conclusive, if the balance of likely costs and benefits justifies it;

- to help people to make better and more informed choices as consumers, investors and voters, there must be improved access to and understanding of information on environmental matters;

- since many of the world's environmental problems are global in scale, international action is necessary to deal with them; and

- since safeguarding the environment can be very costly in the short term, whatever the longer term benefits, it is important to adopt the most cost-effective instruments. In particular, the government will take account of the argument that market-based instruments will often be more efficient and less expensive than regulation in reducing emissions because they allow producers and consumers, rather than regulators, to decide how energy can best and most economically be used.

The United Kingdom is a net exporter of energy, with abundant supplies of oil, gas and coal, as well as existing nuclear generating capacity in place. The United Kingdom is a major supplier of crude oil to OECD countries. It also exports and imports considerable amounts of petroleum products. Concerning natural gas exports, there are at present only limited opportunities to sell new gas into continental markets, but the government will consider applications for waiving the landing requirement on a case-by-case basis.

3. RELEVANT STUDIES

Reports which have contributed to the development of the UK Government's policies and proposals, include:

Department of the Environment, *Climate Change: Report on the United Kingdom's National Programme for Limiting Carbon Dioxide Emissions,* May 1992.

Energy Paper 55: *Renewable energy — the way forward.*

Energy Paper 56: *Prospects for the use of advanced coal-based power generation plant in the United Kingdom.*

Energy Paper 58: *An evaluation of energy-related greenhouse gas emissions and measures to ameliorate them.*

House of Commons *Energy Committee Report: Energy Policy Implications of the Greenhouse Effect;* and government response.

House of Lords Science & Technology Committee Report: *Greenhouse effect;* and government response.

House of Lords Select Committee on the European Communities: *Report on Carbon/Energy Tax;* and government response.

Pezzey, J. (1991): *Imports of Greenhouse Gas Control Strategies on the Competitiveness of the UK Economy:* Report for Department of Trade and Industry.

Work under way and planned for the future includes:

Continuing collaboration in international climate research projects, for example, to improve monitoring, understanding and prediction of the climate, to study the cycling of carbon and other elements in the ocean and related atmospheric changes, to investigate changes in water and energy balance of land following deforestation, and to estimate the CO_2 storage potential of trees and the effects of increased CO_2 on tree growth;

Continuing research on energy efficiency and alternative energy;

Continuing research by government and major public transport operators on transport demand;

Proposed government study on relationship between land use and transport.

United Kingdom
Key Energy and Environmental Data
(1990 data)

Adjusted TPES (M(toe)):	212.15
% Total OECD TPES:	5.15
% Total World TPES:	2.70
Per capita TPES (toe per person):	3.70
Adjusted per capita TFC (toe per person):	2.60
TPES/GDP ratio (toe per US$1 000 1985):	0.40
TFC/GDP ratio (toe per US$1 000 1985):	0.28
Energy-related CO_2 emissions per capita (t CO_2 per person):	10.26
Energy-related CO_2 emissions per unit of GDP (t CO_2 per US$1 000 1985):	1.11
% Total energy-related OECD CO_2 emissions:	5.66
% Total energy-related World CO_2 emissions:	2.73
% Population growth (yearly average growth 1981-1990):	0.21

UNITED STATES

1. OFFICIAL POSITION

The United States believes that a successful climate change policy must be:
- comprehensive, addressing all greenhouse gases, and both their sources and sinks;
- long-term, taking account of the social, economic and environmental consequences and effectiveness of policies over the long term;
- flexible, building on diverse actions which are readily adjustable as knowledge improves through a strong research and development programme; and
- global, designed to involve all nations and incorporate the unique circumstances of each in the development of response strategies.

The United States is undertaking and planning substantial actions to implement this policy. Examples of actions to reduce emissions of greenhouses gases include: amendments to the Clean Air Act which will reduce CO_2 and methane as well as the greenhouse gas precursors such as nitrogen oxides, volatile organic compounds and carbon monoxide; phasing out production and consumption of CFCs and related gases by 31 December 1995 on a schedule which accelerates the requirements of the Montreal Protocol; new initiatives, including expanded appliance efficiency standards and measures to accelerate adoption of energy efficiency and renewable energy technologies; proposed regulations to capture emissions of VOCs and methane from landfills; and continuing to implement a programme to plant one billion trees per year.

The National Energy Strategy (NES) provides for a series of additional actions. It includes actions to: encourage greater use of alternative fuels for transport and to improve vehicle fuel efficiency; improve building and industrial efficiency; encourage integrated resource planning in electricity markets; and increase the availability of energy supply technologies associated with low or minimal net greenhouse gas emissions.

Underlying United States' actions is the largest climate change science and technology research programme among all OECD countries. The United States is now investing over

$1.0 billion annually in its Global Change Research Program to address the critical scientific and economic uncertainties identified by the IPCC.

The United States is committed to seeking a global response to climate change, on a comprehensive and integrated basis, through continued active participation in the Intergovernmental Negotiating Committee (INC) and related scientific, economic and technical assessments conducted by the IPCC. President Bush signed the Convention during the UNCED held in Rio de Janeiro, Brazil, in June 1992. At that conference, President Bush urged a prompt start to action under the Convention, proposing that developed countries prepare and review their strategies for reducing net greenhouse gas emissions by 1 January 1993.

United States' climate change policy is described in detail in several reports, for example those issued in February 1991 by the US Administration: the "National Energy Strategy: First Edition"; "America's Climate Change Strategy — An Action Agenda"; "A Comprehensive Approach to Addressing Potential Climate Change", "National Energy Strategy: One Year Later"; and several papers presented during the Convention negotiations, including "US Views on Climate Change".

2. FACTORS INFLUENCING DECISIONS

The United States supports the need for a response to climate change that is global and comprehensive vis-a-vis: continued research to help resolve the remaining scientific, technical and economic uncertainties and to develop and deploy appropriate technologies; aggressive pursuit of actions now which are already justified on other grounds; and consideration of additional measures as the scientific, economic and technical evidence justifies. The United States strongly endorses the findings of the IPCC report which called on governments to take actions which are:

- beneficial for reasons other than climate change and justifiable in their own right;

- economically efficient and cost-effective, in particular those that use market-based mechanisms;

- able to serve multiple social, economic and environmental purposes;

- flexible and phased, so that they can easily be modified to respond to increased understanding of scientific, technological and economic aspects of climate change;

- compatible with economic growth and the concept of sustainable development;

- administratively practical and effective in terms of application, monitoring and enforcement; and

- reflecting obligations of both industrialised and developing countries, while aware of the special needs of the developing countries, especially in the areas of financing and technology.

The United States also endorses the IPCC's finding that "... there is no single technological quick-fix; a comprehensive strategy addressing all aspects of the problem and reflecting environmental, economic and social costs and benefits is necessary ..." In November 1990, the government ministers at the Second World Climate Conference (SWCC) declared: "We recommend that in the elaboration of response strategies, over time, all greenhouse gases, sources and sinks be considered in the most comprehensive manner possible ...".

3. RELEVANT STUDIES

A. Completed Studies

DOE: *National Energy Strategy: First Edition 1991/1992,* February 1991.

DOE: *National Energy Strategy: One Year Later,* February 1992.

The White House: *America's Climate Change Strategy: An Action Agenda,* February 1991.

DOS: *US Views on Climate Change,* February 1992.

DOC: *Economic Effects of Using Carbon Taxes to Reduce Carbon Dioxide Emissions in Major OECD Countries,* January 1992.

DOE: *Limiting Net Greenhouse Gas Emissions in the United States,* September 1991.

EPA/DOS: *US Efforts to Address Global Climate Change:* Report to Congress, February 1991.

Interagency Task Force on the Comprehensive Approach to Climate Change (chaired by DOJ): *A Comprehensive Approach to Addressing Potential Climate Change,* February 1991.

EPA: *The Potential Effects of Climate Change on the United States,* J. Smith and D. Tirpak eds., 1989.

EPA: *Costs and Benefits of Phasing Out Production of CFCs and Halons in the United States,* 1989.

EPA: *Estimating Regional Anthropogenic Emissions of Greenhouse Gases,* Ahuja, D.R., September 1990.

DOE: *Energy Efficiency: How Far Can We Go?,* Carlsmith, R.G. et al, 1989.

DOE: *The Potential of Renewable Energy: An Interlaboratory Analytic Paper,* Solar Energy Research Institute, 1989.

DOE: *National Energy Strategy,* Interim Report, April 1990.

DOE: *A Fossil Energy Perspective on Global Climate Change,* January 1990.

DOE: *Global Climate Trends and Greenhouse Gas Data: Federal Activities in Data Collection, Archiving and Dissemination,* June 1990.

DOE: *The Economics of Long-Term Global Climate Change:* Report of an Interagency Taskforce headed by the CEA, September 1990.

DOE: *Compendium of Options for Government Policy to Encourage Private Sector Responses to Potential Climate Change,* A Report to the Congress of the United States, October 1989.

DOE: *The Prospect of Solving the CO_2 Problem Through Global Reforestation,* Marland, G., 1988.

EPA: *Interim Report of Long-Term Energy Scenarios for Developing Countries,* Lawrence Berkeley Laboratories, August 1990.

EPA: *Report of the IPCC Energy and Industry Subgroup,* January 1991.

AID: *Greenhouse Gas Emissions and the Developing Countries: Strategic Options and the USAID Response.* A Report to Congress, July 1990.

EPA: *Policy Options for Stabilizing Global Climate,* December 1991.

EPA: *Cooling our Communities,* April 1992.

CBO: *Carbon Charges as a Response to Global Warming: The Effects of Taxing Fossil Fuels,* August 1990.

OTA: *Changing by Degrees: Steps to Reduce Greenhouse Gases,* February 1991.

NOAA: *Economics and Global Change:* New Haven Workshop, May 1990.

B. Conference and IPCC Proceedings

EPA: Proceedings of North American Conference on Forestry Responses to Climate Change, September 1990.

EPA: Proceedings of IPCC Tropical Forestry and Climate Change Workshop, July 1990.

EPA: Proceedings of IPCC Agriculture Workshop, September 1990 (published by the EPA for the IPCC).

EPA: Proceedings of 1st North American Forestry and Climate Change Conference, November 1990 (published by the EPA for the IPCC).

EPA: Sea Level Rise Report: Changing Climate and the Coast, 1990 (Published by the EPA for the IPCC).

United States

Key Energy and Environment Data
(1990 data)

TPES:	1 871.20
% Total OECD TPES:	45.39
% Total World TPES:	23.82
Per capita TPES (toe per person):	7.44
Adjusted per capita TFC (toe per person):	5.33
TPES/GDP ratio (toe per US$1 000 1985):	0.41
TFC/GDP ratio (toe per US$1 000 1985):	0.29
Energy-related CO_2 emissions per capita (t CO_2 per person):	19.97
Energy-related CO_2 emissions per unit of GDP (t CO_2 per US$1 000 1985):	1.09
% Total energy-related OECD CO_2 emissions:	48.27
% Total energy-related World CO_2 emissions:	23.27
% Population growth (yearly average growth 1981-1990):	0.97

THE EUROPEAN COMMUNITY

1. OFFICIAL POSITION

In May 1992, the Commission adopted, subsequent to its Communication of 14 October 1991 and in response to the request made by the joint energy/environment Council of December 1991, a proposal to be made to the Council on a Community Strategy to reduce CO_2 emissions and to improve energy efficiency. The objective of this strategy is to implement the Council decision of October 1990 to stabilize CO_2 emissions in the Community in 2000 at their 1990 levels, through reduced energy demand, increased energy efficiency and modification in the energy sources used[1]. It should be implemented through a package of measures in the energy field and fiscal measures. The measures have been drafted in such a way that apart from realising the key objective, they will also bring about other so-called "no-regret" benefits.

The energy measures are mainly focussed on an improved development of alternative non-polluting energy sources, on the strengthening of energy efficiency in the industrial and housing sector, on the promotion of an improved infrastructure for the transport and on the optimisation of energy conservation techniques.

The measures will be implemented in the programmes SAVE, ALTENER and THERMIE. Together, they are expected to reduce the anticipated 12% increase of CO_2 emissions by about 5.5% up to the year 2000.

In the context of SAVE programme on energy efficiency adopted on 29 October 1991, the Commission has proposed a framework Directive foreseeing specific action in the buildings, transport and industry sectors. These actions focus on information of the consumers through actions on energy certification of buildings, and on energy audits in industry as well as on development of a more responsible behaviour of the same consumers (real billing of energy

---

1. Emissions of CO_2, which come primarily from the use of fossil fuels, are recognised as the principle cause of global warming since they account for 61% of greenhouse gases. A modification of overall climate equilibrium would have negative impact on ecosystems, on the habitats and on species. At present, it is the industrialised world which is mainly responsible for the CO_2 emissions. The share of the European Community is 13%, that of the United States 23% and Japan 5%. Central and Eastern European countries and the CIS contribute about 25%.

consumption, thermal insulation, third-party financing and periodical inspections of certain equipment, as for example, boilers and cars).

As energy consumption is the main source for the emission of greenhouse gases, an increased use of renewable energy sources has become a priority.

The ALTENER programme proposed by the Commission to the Council focusses on:
- a further penetration of renewable energy sources on the market;
- economic and financial actions;
- information; and
- co-operation with countries which are not members of the European Community.

Harmonization and standardization of legislation in Member States will be the actions of priority.

Quantified objectives are proposed by the Commission for 2005:
- double the share of renewables in the total energy package from its current 4% to 8%;
- triple its share in electricity generation; and
- cover 5% of transport fuels by biofuels.

Development and introduction of new technologies has been promoted by the Community over the last 15 years through financial support. Indeed, more than 2 700 projects have received such support involving a budget of ECU 1.7 billion.

Today, this is mainly done under the THERMIE project. After consultations with the Member states, conclusion was reached that re-orientation of the programme towards projects having a direct impact on CO_2 emissions would be desirable and indeed is possible. The next tender round will, therefore, be directed to that end.

The other 6.5% reduction in CO_2 emissions required to realise the stabilization target could, in the opinion of the Commission, best be covered by national programmes and a Community CO_2/energy tax. A Directive to that end has been submitted to the Council. The 50% energy part of the tax should, by its impact on energy prices, stimulate the more efficient use of available energy sources while the 50% CO_2 part should, in addition, stimulate the switch towards less or non-CO_2-containing fuels. The tax would be a Community tax to be levied by Member states.

With the exception of renewables, all energy sources should be taxed. Large hydropower installations should, however, be included. The tax should be an excise-like product tax; primary energy sources will be taxed at the level of consumption.

The level of the tax, which will be a minimum level that could be raised by the Member states wishing to do so, will be US$3 (ECU 2.81 per tonne of CO_2 emitted by fossil fuels and ECU 0.21 per gigagoule) on 1 January 1993 to be increased by $1 a year up to US$10 in 2000.

The principle of revenue neutrality, that is to say, the tax should not result in an overall increase in statutory contributions and charges, should be respected.

It is also important that the tax should not jeopardise the international competitive position of the industry sector in the Community. Therefore, the imposition of tax has been made subject to the condition that other OECD countries take actions resulting in a similar financial charge as the one resulting from the measures foreseen in the Strategy. Furthermore, reductions or exemptions from the tax for companies with high energy consumption and operating on the world market are, under certain conditions, foreseen. Finally, in this context, reductions on tax payments are possible for all industries for investments in energy efficiency or in emissions' reduction in the Community as well as in countries which are not members of the European Community over a period of three years.

A proposal for a Directive on a monitoring mechanism to follow up the national implementation plans and programmes of the Member states as well as the efficiency of the Community measures has been included in the strategy proposal. Apart from the CO_2 emissions, it would also cover other greenhouse gas emissions as well as the equitable sharing of the burdens resulting from the implementation of the strategy between the Member states. Indeed, with regard to the latter, financial assistance to Member states that would have difficulties in financing the required investments to adapt to the new situation is foreseen.

The macro-economic as well as the impact on households could be kept rather modest if the additional tax will be compensated for by lowering other taxes or charges and the introduction will be done in a gradual and predictable way.

Finally, it is important to mention in this context that the Convention on Climate Change has been signed in Rio during UNCED by the Community and its Member states. Although, the commitments reflected in this convention are not as far-reaching as those reflected above, they are going in the same direction. In a formal declaration, the Community desired a prompt start of the implementation of the Convention and re-iterated the necessity of re-inforcement of the commitments entered into, by inter alia the adoption of an additional Protocol.

2. RELEVANT STUDIES

Communication to the Council on "The Greenhouse Effect and the Community" (COM(88)656 final).

Communication to the Council on "Energy and the Environment" (COM(89)369 final).

Communication to the Council on Community Policy Targets on the Greenhouse Issue (SEC(90)496 final).

Etude sur le CO₂ Crash Programme, Mars 1991, rapport pour la Commission des Communautés Européennes, DG XII.

Report of the Working Group on the Use of Economic and Fiscal Instruments in EC Environmental Policy, DG XI/185/90, September 1990.

The Economics of Policies to Stabilize or Reduce Greenhouse Gas Emissions: The Case of CO_2 (draft 11.10.90: II/335/90-EN).

"CO_2 Targets and Burden Sharing", *Energy in Europe no. 16*.

Resolution of the Council on "The Greenhouse effect and the Community" (89/C183/03).

Energy for a new Century: The European Perspectives, July 1990.

The European Community

Key Energy and Environmental Data
(1990 data)

Adjusted TPES (Mtoe):	1 246.98
% Total OECD TPES:	30.25
% Total World TPES:	15.87
Per capita TPES (toe per person):	3.63
Adjusted per capita TFC (toe per person):	2.54
TPES/GDP ratio (toe per US$1 000 1985):	0.42
TFC/GDP ratio (toe per US$1 000 1985):	0.29
Energy-related CO_2 emissions per capita (t CO_2 per person):	127.55
Energy-related CO_2 emissions per unit of GDP (t CO_2 per US$1 000 1985):	16.24
% Total energy-related OECD CO_2 emissions:	30.58
% Total energy-related World CO_2 emissions:	14.74

SECTION 3

This Section gives a brief description of energy-related climate change policy initiatives of major non-OECD emitters of greenhouse gases. The effect on the relative rankings of countries of the breakup of the USSR is to put three of the resulting republics (Russia, Ukraine, and Kazakhstan) on the list of the top non-OECD energy-related CO_2 emitters. Hence, in this section the three republics' country profiles have replaced the one for the USSR which appeared in earlier updates to the binder[1].

Some of these country descriptions have been reviewed by a government ministry or institute which has been active in either the INC or the Intergovernmental Panel on Climate Change. Those for which no input was received from the country are presented as the "Apparent Position". Those which were reviewed by some part of the country's government are presented as "Position".

In the Key Energy and Environmental Data sub-sections at the end of each country profile, it should be noted that the TPES reported and the derived CO_2 emissions are calculated using the IPCC/OECD methodology (see Section 1 for an explanation). Population estimates and the growth rates derived from them are taken from the *United Nations Monthly Bulletin of Statistics,* February 1992.

---

1. For these descriptions we have relied heavily upon two publications: *Commonwealth of Independent States Country Report, N°.1,* The Economist Intelligence Unit, London, 1992 and *International Energy Outlook 1992,* Energy Information Administration, Washington D.C, April 1992.

BRAZIL

1. POSITION

Brazil organised and hosted UNCED held in Rio de Janeiro from 3-14 June 1992. Brazil's President was the first to sign the Convention when it was opened for signature there. Brazil has also been very active in both the INC and the IPCC. In these fora, the government has stated that the issue of global warming cannot be separated from wider development issues and the relationship between developed and developing countries. Brazil has insisted upon the sovereign right of countries to exploit their own resources. It has also maintained that "all obligations and commitments to be taken by developing countries are conditioned to and dependent upon the provision of new and additional financial resources and the transfer of technology on a non-commercial and preferential basis". Brazil signed the June 1991 Beijing Declaration.

Brazil has implemented policies which have already produced results in reducing the rate of deforestation by at least a factor of two in recent years, and in introducing alcohol produced from sugar cane as a fuel replacement to gasoline, to an extent such that their annual consumption is of the same order of magnitude. Brazil and the United States have agreed on a joint project to carry out a comprehensive greenhouse gas emissions inventory.

2. FACTORS INFLUENCING DECISIONS

Brazil is the most heavily indebted nation in the world and is experiencing continued high population growth rates.

The contribution of Brazil to net greenhouse gas emissions is apparently still dominated by deforestation, as a result of its very large forests and of the fact that two-thirds of the energy is produced by renewable sources, notably hydroelectrical.

In energy-related carbon dioxide emissions Brazil ranks eighth in the non-OECD (and seventeenth globally). Brazil is the world's seventh largest major contributor to net greenhouse gas emissions which reflects the effects of deforestation.

Brazil is highly dependent on petroleum even with the world's largest alcohol fuels use and its hydroelectricity generation. Brazil's own analyses indicate considerable scope for further energy efficiency improvements and fuel-switching.

Brazil

Key Energy and Environmental Data
(1990 data)

Adjusted TPES (Mtoe):	91.68
% Non-OECD TPES:	2.46
% Total World TPES:	1.18
Per Capita TPES (toe per person):	0.61
Per Capita TFC (toe per person):	0.53
TPES/GDP ratio (toe per US$1 000 1985):	0.36
TFC/GDP ratio (toe per US$1 000 1985):	0.32
Energy-related CO_2 emissions per capita (t CO_2 per person):	1.48
Energy-related CO_2 emissions per unit GDP (t CO_2 per US$1 000 1985):	0.88
% Total energy-related Non-OECD CO_2 emissions:	2.00
% Total energy-related World CO_2 emissions:	1.03
% Population growth (yearly average growth 1981-1990):	2.36

CHINA

1. POSITION

China attended UNCED in Rio in June 1992 and signed the Convention. China participated actively in the United Nations' negotiations on climate change and in the IPCC, co-chairing its Energy and Industry Subgroup of the Working Group III.

China hosted the Ministerial Conference of Developing Countries on Environment and Development in Beijing, 14-19 June 1991 resulting in the Beijing Declaration, the text of which is at the end of this section, which affirms participation in the international negotiating process on the basis of differentiated responsibilities between developing and developed countries and that developed countries must provide adequate technology transfer and financing.

Throughout the negotiations China stressed that there are major scientific uncertainties and lack of agreement on which measures should be taken but that climate change is a common concern of mankind and that there is a need for effective international co-operation based on equity. China supported having a convention which set general principles and obligations, "thereby preparing the ground for the eventual establishment of a legal regime". China repeatedly called for consideration of the need for developing countries to continue developing their economies to meet basic needs of their populations and of the fact that most developing countries' energy consumption per capita is low.

China has insisted that the developed world needs to recognise its responsibility for "human-induced climate change and their obligation in addressing it". This means that different timeframes would need to be set given the differences economically and technologically.

China has been pursuing economic development as its central task and now enjoys sustained economic growth and improving living standards. China takes account of environmental protection in the course of economic development. Environmental protection is a basic state policy. China has devised a strategy of synchronised planning, implementation and development in terms of economic development, urban and rural construction, and environmental protection, and has adopted three major principles, i.e. to put prevention first, to hold those who cause pollution responsible for cleaning up and to strengthen environmental control and management. They have improved the legal system for environmental protection and set up relevant organs at various levels and an interministerial co-ordination agency at the national level. They have pooled the efforts of various quarters to clean up urban environments and to prevent and control industrial pollution. They have conducted extensive education in environmental protection to awaken the whole nation to its importance.

While China has made no specific commitment to greenhouse gas emissions limitation, it accepts in principle that emissions targets should be set. Such a Chinese commitment could only be made if there was a transfer of technology on favourable terms, and if there was additional financial assistance for such transfer from developed countries.

China is participating in a least-cost greenhouse gas emissions reduction project with the assistance of UNDP and the Global Environment Fund.

2. FACTORS INFLUENCING DECISIONS

China occupies third place in the world, following the United States and the Russian Federation, in energy production and consumption. (Until its dissolution, the Soviet Union was in second place.) In 1989, total primary energy production reached 697 Mtoe. From 1980 to 1989, the average annual growth rate of energy production was about 6.8%.

China is one of the few countries in the world which uses coal as the principal energy source. China is highly dependent on coal (around 76% in 1988) for direct consumption and for the generation of electricity. The country has an ambitious industrialisation programme based in large part on the use of its indigenous fuels. According to ESCAP, coal consumption is expected to more than double between 1988 and 2010 (490 Mtoe in 1988, 707 Mtoe in 2000 and 1 003 Mtoe in 2010).

Energy consumed in daily life by urban residents (about 20% of China's population) now is almost exclusively commercial energy. But in rural areas, commercial energy accounts for only one-quarter of the energy consumption. A large quantity of biomass energy has to be consumed. At present, about 200 million rural residents live without electricity. Firewood, straw and stalks are the main sources of biomass energy.

Studies show that coal and other forms of energy are used very inefficiently. Programmes have been initiated to increase coal combustion efficiency; however, the pricing of coal gives little incentive for energy saving. It is estimated that the marginal cost of coal is about twice as much as the price received by state-owned mines.

China's population is about 1.1 billion, the highest in the world. China contributes 11% of world CO_2 emissions, which ties it with the Russian Federation for second. (Until its dissolution, the Soviet Union was in second place.)

China

Key Energy and Environmental Data
(1990 data)

Adjusted TPES (Mtoe):	662.00
% Non-OECD TPES:	17.73
% Total World TPES:	8.52
Per Capita TPES (toe per person):	0.58
Per Capita TFC (toe per person):	0.44
TPES/GDP ratio (toe per US$1 000 1985):	1.59
TFC/GDP ratio (toe per US$1 000 1985):	1.20
Energy-related CO_2 emissions per capita (t CO_2 per person):	2.11
Energy-related CO_2 emissions per unit GDP (t CO_2 per US$1 000 1985):	5.78
% Total energy-related Non-OECD CO_2 emissions:	21.49
% Total energy-related World CO_2 emissions:	11.13
% Population growth (yearly average growth 1981-1990):	1.44

CZECH AND SLOVAK FEDERAL REPUBLIC (CSFR)

1. APPARENT POSITION

The CSFR attended the negotiations on the convention on climate change as well as the UNCED. The Environment Minister stated at UNCED in his address to the Plenary that the CSFR would sign the Convention "after completing the necessary internal procedures".

In the recent governmental and economic restructuring of the country, new environmental laws covering conventional air pollutants and other energy-related environmental problems have been adopted. Strict emission limits with fines for non-compliance are in effect and fees for air pollutant emissions are being phased in.

The former government of Czechoslovakia had signed the United Nations Economic Commission for Europe's (UNECE) Convention on Long-Range Transboundary Air Pollution and its two protocols, on SO_2 and NO_x emissions. More recently, the new government has been active with the UNECE in organising the first pan-European Conference of environmental ministers which established a framework of a future Environmental Programme for Europe oriented, *inter alia,* on cleaning up the most polluted industrial areas.

2. FACTORS INFLUENCING DECISIONS

The entire economy is undergoing restructuring and modernising. The transition will take time and many policy areas must be dealt with simultaneously. Restructuring is expected to bring greater efficiency of energy use leading in turn to reduced environmental impacts, including emissions of greenhouse gases. However, uncertainty about changes in demand leaves a question about whether expectations for reductions, especially those due to economic recession, might be offset with somewhat longer-term increased demand as the standard of living improves.

The CSFR is highly dependent on both hard coal and lignite. Coal represents almost 57% of TPES. Of that, three-quarters is lignite. Coal is used extensively in all sectors except transport. Government projections show lignite production dropping 40% between 1990 and 2005 and hard coal production dropping 50%. In the residential sector, it is expected that lignite will be replaced by natural gas and light fuel oil.

The CSFR is also highly industrialised. A very high percentage of its industry is energy-intensive. The iron and steel industry, alone, uses 47% of total hard coal consumption. Combustion efficiencies are considered quite low in all sectors. There is a strong anti-coal feeling in the country because of the associated environmental damage, especially that caused by the emissions of sulphur dioxide associated with coal combustion. The clear and present problem of very high and concentrated SO_2 emissions from coal combustion are recognised as the first order of priority in the near term for the CSFR.

Czech and Slovak Federal Republic

Key Energy and Environmental Data
(1990 data)

Adjusted TPES (Mtoe):	68.49
% Non-OECD TPES:	1.83
% Total World TPES:	0.88
Per Capita TPES (toe per person):	4.37
Per Capita TFC (toe per person):	2.91
TPES/GDP ratio (toe per US$1 000 1985):	1.69
TFC/GDP ratio (toe per US$1 000 1985):	1.12
Energy-related CO_2 emissions per capita (t CO_2 per person):	13.65
Energy-related CO_2 emissions per unit GDP (t CO_2 per US$1 000 1985):	5.26
% Total energy-related Non-OECD CO_2 emissions:	1.92
% Total energy-related World CO_2 emissions:	0.99
% Population growth (yearly average growth 1981-1990):	0.25

INDIA

1. POSITION

India attended UNCED in Rio and signed the Convention. India has also been active in the INC and the IPCC as a strong proponent of developing countries' needs and positions. India is a signatory to the June 1991 Beijing Declaration. India believes that feasible measures can be considered "in accordance with their national development plans, priorities and objectives". India has also stressed the need for a "Climate Fund" to help developing countries "adapt to and mitigate the adverse effects of climate change. . .".

India worked hard to have included in the Convention statements that "the largest part of the current emission of pollutants into the environment originates in developed countries, and recognising therefore that these countries have the main responsibility for combating such pollution". India has further emphasized the importance of developing country access to advanced and alternate technologies on a cost-free basis wherever these are justified for global environmental reasons.

India is participating in an Asian Development Bank project to develop a regional strategy to address the impacts of climate change which includes country studies on potential impacts, possible policy options to reduce emissions and to adapt to climate change, and national response strategies.

India has recently agreed to sign the Montreal Protocol at which point it will have committed itself to controlling its CFC consumption in approximately ten years.

2. FACTORS INFLUENCING DECISIONS

India is the second most populous nation in the world with the present population growth rate measuring over 2% per year. Its GDP has grown at a rate of 5% per year and its future projections predict a further acceleration in this rate.

TPES has grown at a rate of 6.0% per year between 1973 and 1989. Energy use per capita, at 0.21 toe per person, is very low. Energy intensity has been increasing with a rising rate of increase, largely due to the fast pace of industrialisation. Growth in commercial energy consumption has been around 7% per year with conversion and distribution losses around 30% of the gross availability of commercial energy.

India is highly dependent on coal, which represents almost 57% of TPES. This dependence will continue even though natural gas and other energy sources are slowly increasing their shares. Coal consumption is expected to go from 91 Mtoe in 1988 to 160 Mtoe in 2000 and 280 Mtoe in 2005, continuing the recent trend of growth of 11% per year. Oil, which represented a further 33.7% of TPES in 1989, will rise from 48 Mtoe in 1988 to 82 Mtoe in 2000 to 118 Mtoe in 2005. Natural gas use has been growing at a rate of almost 15% per year. Biomass fuels have a large share in fulfilling energy demands in India, especially in the rural sector.

India

Key Energy and Environmental Data
(1990 data)

Adjusted TPES (Mtoe):	178.08
% Non-OECD TPES:	4.77
% Total World TPES:	2.29
Per Capita TPES (toe per person):	0.22
Per Capita TFC (toe per person):	0.15
TPES/GDP ratio (toe per US$1 000 1985):	0.63
TFC/GDP ratio (toe per US$1 000 1985):	0.44
Energy-related CO_2 emissions per capita (t CO_2 per person):	0.72
Energy-related CO_2 emissions per unit GDP (t CO_2 per US$1 000 1985):	2.10
% Total energy-related Non-OECD CO_2 emissions:	5.32
% Total energy-related World CO_2 emissions:	2.75
% Population growth (yearly average growth 1981-1990):	2.21

KAZAKHSTAN

1. APPARENT POSITION

After the dissolution of the USSR, Kazakhstan was given a seat in the INC but was not represented. However, Kazakhstan did attend UNCED where it signed the Convention.

2. FACTORS INFLUENCING DECISIONS

Kazakhstan only became independent when the USSR was dissolved on 1 January 1992 as it did not declare itself independent before that. Its population is approximately half Kazakhs and half Russian and inter-ethnic strife is not uncommon. Rapid movement to a market economy was the course decided for Kazakhstan's transition. However the freeing of prices in 1992 resulted in the coal miners going on strike. Social safety nets are being devised. Kazakhstan has stated that it would like to become a member of the European Community.

Kazakhstan is a major producer of energy, especially coal, but also oil and gas. Significant amounts of coal are found in the basins in the north-eastern part of the country, near West Siberia, while large reserves of oil and natural gas are located in the north west near the Caspian Sea. Of particular interest is the Tengiz oil field, which was discovered in 1979 and may contain upwards of 20 billion barrels of oil. Development of Tengiz has been impeded by, among other things, Soviet drilling and production equipment which is poorly-suited to handling the highly corrosive sour crudes and abnormally high downhole pressures of Tengiz. Kazakhstan recently agreed with Elf-Aquitaine to develop oil fields in the western part of the country. Other companies are trying to conclude agreements to develop the Tengiz field and other oil fields.

Overall, Kazakhstan produces around 26.9 million tons of oil and 250 billion cubic feet of natural gas. It consumes around 18.7 million tons of oil and 400 billion cubic feet of gas. Kazakhstan is the most energy-intensive former republic of the USSR.

Kazakhstan

Key Energy and Environment Data
(1990 data)

Adjusted TPES (Mtoe):	69.42
% Non-OECD TPES:	1.86
% Total World TPES:	0.89
Per Capita TPES (toe per person):	4.15
Per Capita TFC (toe per person):	N.A.
TPES/GDP ratio (toe per US$1 000 1985):	N.A.
TFC/GDP ratio (toe per US$1 000 1985):	N.A.
Energy-related CO_2 emissions per capita (t CO_2 per person):	13.86
Energy-related CO_2 emissions per unit GDP (t CO_2 per US$1 000 1985):	N.A.
% Total energy-related Non-OECD CO_2 emissions:	2.08
% Total energy-related World CO_2 emissions:	1.08
% Population growth (yearly average growth 1981-1990):	N.A.

MEXICO

1. APPARENT POSITION

Mexico attended UNCED in Rio in June 1992 where it signed the Convention.

Mexico has been very active in both the INC (providing a co-chairman to its Working Group I) and the IPCC. It signed the June 1991 Beijing Declaration. In the negotiations Mexico maintained that the Convention should provide preferential access to environmental technologies, such as those for natural gas conversion, energy technologies for sustainable agriculture, or pollution control. At the INC, Mexico proposed a "technology facility" to be under the control of the Conference of the Parties to identify appropriate technologies and organise competition for acquisition of those technologies.

Mexico is giving more attention to environmental issues and in March 1991, Mexico City released a comprehensive plan to reduce air pollution in the city. Recently, Mexico agreed to a joint project with the United States to inventory greenhouse gas emissions and assess technical and policy options to limit or reduce greenhouse gases.

2. FACTORS INFLUENCING DECISIONS

Mexico is the second most indebted nation in the world behind Brazil with a total foreign debt of $107 billion in 1989. Since the oil price decrease in 1986 and the earthquake in Mexico City in 1985, the economy has stagnated. Mexico is highly dependent on oil for export revenues; oil's share dropped from over 65% of export revenues in 1985 to less than one-third in 1989. Nevertheless, Mexico remains the world's fifth largest oil exporter.

Mexico has a large energy-intensive industrial sector based on petrochemicals and iron and steel. Mexico is currently negotiating with the United States and Canada for a North American free trade pact.

Almost 70% of TPES is oil with a further 22% being natural gas. Coal represents a very small share. All end-use sectors are highly dependent on oil (45% in industry, 100% in transport and 77% in the residential/commercial sector). Energy intensity is quite high and there is scope for efficiency improvements throughout the entire energy system.

Mexico

Key Energy and Environmental Data
(1990 data)

Adjusted TPES (Mtoe):	118.95
% Non-OECD TPES:	3.19
% Total World TPES:	1.53
Per Capita TPES (toe per person):	1.38
Per Capita TFC (toe per person):	1.03
TPES/GDP ratio (toe per US$1 000 1985):	0.60
TFC/GDP ratio (toe per US$1 000 1985):	0.45
Energy-related CO_2 emissions per capita (t CO_2 per person):	3.72
Energy-related CO_2 emissions per unit GDP (t CO_2 per US$1 000 1985):	1.63
% Total energy-related Non-OECD CO_2 emissions:	2.87
% Total energy-related World CO_2 emissions:	1.49
% Population growth (yearly average growth 1981-1990):	2.30

POLAND

1. POSITION

Poland attended UNCED in Rio in June 1992 and signed the Convention. In these discussions, it helped to articulate the perspective and particular situation of the economies in transition to market economies and to gain separate treatment in the Convention. Poland has also been active at the INC and in the IPCC's efforts to establish a methodology for estimating greenhouse gas emissions on a comparable basis.

The Polish Government first officially defined its position on greenhouse gases policy commitments during the Nordwijk Ministerial Conference (1989) where the Minister of Environmental Protection, Natural Resources and Forestry stated that the entire activity being actually undertaken in Poland in the field of environmental protection should have the effect in the next 10 to 15 years inter alia of reaching the emission levels of 1988 in the period 2005 to 2010. Also during the Second World Climate Conference in Geneva (1990), the Polish representative repeated the above position declaring that Poland would be aiming at achieving, by the year 2000, the stabilization of the CO_2 emissions at the level of the years 1988 to 1989 and that the efficiency of these efforts depended on the progress in the stabilization of the Polish economy and the ability to have access to advanced and environmentally-sound technologies.

Poland has a co-operative study in process with the United States on emissions inventory and emissions reduction options and is a participant country in a UNEP/GEF project on greenhouse gas sources and sinks.

2. FACTORS INFLUENCING DECISIONS

Poland is highly dependent on coal both for energy consumption and for trade. In 1989, over 78% of TPES was coal. In 1990 Poland was also the third largest non-OECD coal exporter and sixth in the world. Poland is extracting and consuming ever increasing shares of lignite, which can adversely affect air pollution emissions.

Poland is modernising its entire economy but the transition is slow and expensive. High foreign debt is a barrier to the modernisation process, preventing closures of very old and polluting plants. Industry has traditionally been highly energy-intensive, using domestic coal for its fuel supply. Intense coal use is recognised to have been a major cause of environmental degradation. In 1990, the Ministry of Environmental Protection estimated that it would take $20 billion to clean up existing environmental problems over the next 20 years. In 1991, the Polish Parliament passed the "National Environmental Policy". It is being implemented by "environmentally-sound restructuring of the economy", letting energy prices rise, modernisation and switching to cleaner fuels which will undoubtedly reduce carbon dioxide emissions although investment costs will be high. However, as the economy develops, total energy use and emissions could rise as the demand for consumer products rises.

Poland

Key Energy and Environmental Data
(1990 data)

Adjusted TPES (Mtoe):	98.25
% Non-OECD TPES:	2.63
% Total World TPES:	1.26
Per Capita TPES (toe per person):	2.58
Per Capita TFC (toe per person):	1.73
TPES/GDP ratio (toe per US$1 000 1985):	1.41
TFC/GDP ratio (toe per US$1 000 1985):	0.95
Energy-related CO_2 emissions per capita (t CO_2 per person):	9.40
Energy-related CO_2 emissions per unit GDP (t CO_2 per US$1 000 1985):	5.15
% Total energy-related Non-OECD CO_2 emissions:	3.20
% Total energy-related World CO_2 emissions:	1.66
% Population growth (yearly average growth 1981-1990):	0.71

REPUBLIC OF SOUTH KOREA

1. POSITION

Korea has been active in the INC and attended the UNCED in Rio in June 1992 where it signed the Convention. In his address to the UNCED, the Prime Minister proposed "joint efforts in north-east Asia to establish a regional institution for environmental co-operation", noting the present void in the region.

Previous Korean positions at the INC helped to shape the form of the final Convention. Notably, at the first session of the INC, the Ambassador stated that "the goal of the convention should not be to establish specific guidelines for all countries, but rather, to encourage their voluntary efforts . . ." Quotas "should be done in separate protocols rather than in the Convention". The statement continued to stress that consideration needs to be given to the specific economic and technical capabilities of individual countries under the principle of equity "and the common, but differentiated, responsibilities of the countries". The Convention would also need to ensure special treatment to developing countries for financial assistance "and the preferential transfer of environmental technology".

At the third session of the INC, the Korean statement amplified that some countries "need to maintain rather fast economic growth, at least for some time to come". For countries in the process of industrialisation, these "are the ones that are subject to high sacrifices to climate change convention commitments, at least until they manage to restructure their energy sectors and economies". Korea also stresses that, although some countries may phase out energy-intensive industries, "some other countries might have to fill the gap, often less efficiently. From a global point of view. . . those countries burdened with heavy industries . . . should be credited in one way or the other, at least on a transitionary basis".

In February 1992, Korea acceded to the Montreal Protocol on CFCs.

2. FACTORS INFLUENCING DECISIONS

Korea is highly dependent on imported coal and oil for its rapid economic development. Over 70% of total energy is imported. It has nine nuclear plants and plans for nine more in the 1990s.

Between 1973 and 1989, TPES grew at an annual rate of 7.7% while total GDP grew 8.5%. In recent years, however, energy consumption has started to grow faster than GDP mainly due to the continued expansion of basic energy-intensive industries and growth in income. The increase in per capita income has brought a rapid increase in demand for vehicles and electronics, although still below the level of most industrialised countries.

During the 1970s economic growth was based on energy-intensive industries such as steel, cement and petrochemicals. Overall energy intensity has declined from 0.76 in 1980 to 0.65 in 1989, with some fluctuations paralleling industrial restructuring.

Korea is in the process of strengthening its environmental standards.

Republic of South Korea

Key Energy and Environmental Data
(1990 data)

Adjusted TPES (Mtoe):	93.99
% Non-OECD TPES:	2.52
% Total World TPES:	1.21
Per Capita TPES (toe per person):	2.19
Per Capita TFC (toe per person):	1.68
TPES/GDP ratio (toe per US$1 000 1985):	0.62
TFC/GDP ratio (toe per US$1 000 1985):	0.48
Energy-related CO_2 emissions per capita (t CO_2 per person):	5.74
Energy-related CO_2 emissions per unit GDP (t CO_2 per US$1 000 1985):	1.63
% Total energy-related Non-OECD CO_2 emissions:	2.20
% Total energy-related World CO_2 emissions:	1.14
% Population growth (yearly average growth 1981-1990):	1.17

ROMANIA

1. APPARENT POSITION

Romania attended the UNCED in Rio in June 1992 where it signed the Convention. The speech to UNCED by the President of Romania was the country's first official statement on national, transboundary and global environmental problems, noting that "problems of environment rank high on Romania's priority list". Furthermore, the Minister of Environment of Romania announced at UNCED that Romania expected to be able to have CO_2 emissions in the year 2000 below those of 1989.

Romania has completed several studies on impacts which it plans to continue and has a preliminary assessment of greenhouse gas emissions and sinks under way.

2. FACTORS INFLUENCING DECISIONS

The country is in a state of transition with the economy in a steady decline since 1986. In the past year, since the change in government, the economy has virtually collapsed.

Romania is about 75% self-sufficient in energy. Of non-OECD countries, it is one of the highest consumers of natural gas in terms of the share of TPES (44.5%). Under the previous regime, its economic policy was to develop energy-intensive industries (machinery and transport equipment). In 1988, Romania ranked second in exporting oil industry equipment.

Canada has recently extended new credits to complete the CANDU-style nuclear reactor. Nuclear power is expected to provide almost 7% of TPES by the year 2000 and 21% in 2010.

Romania

Key Energy and Environmental Data
(1990 data)

Adjusted TPES (Mtoe):	58.89
% Non-OECD TPES:	1.60
% Total World TPES:	0.77
Per Capita TPES (toe per person):	2.58
Per Capita TFC (toe per person):	1.84
TPES/GDP ratio (toe per US$1 000 1985):	N.A.
TFC/GDP ratio (toe per US$1 000 1985):	N.A.
Energy-related CO_2 emissions per capita (t CO_2 per person):	7.27
Energy-related CO_2 emissions per unit GDP (t CO_2 per US$1 000 1985):	N.A.
% Total energy-related Non-OECD CO_2 emissions:	1.51
% Total energy-related World CO_2 emissions:	0.78
% Population growth (yearly average growth 1981-1990):	0.42

RUSSIAN FEDERATION

1. POSITION

Until its dissolution, the USSR was active in both the INC and the IPCC and chaired the IPCC's Working Group II on socio-economic impacts. Since the dissolution, the Russian Federation has taken the USSR's seat at the INC (and other independent republics began to attend under their own state flags). The Soviet Union's consistent position was that more scientific study needed to be undertaken on the extent and timing of global climate changes before specific targets could be set. While the Russian Federation attended UNCED in Rio in 1992 at which it signed the Convention, the environmental preoccupations of the country lie in clearing up their nuclear reactor safety problems and cleaning up after the years of environmental destruction. Nevertheless it has completed a preliminary inventory of CO_2 and CH_4 emissions.

2. FACTORS INFLUENCING DECISIONS

The Russian Federation ties with China for second in the world in total energy-related carbon dioxide emissions. The country is currently in the midst of political, administrative and economic restructuring and the final outcome of the new federal system is still to be determined. Output is falling and there is a severe shortage of investment capital.

By itself, Russia was the world's largest oil producer in 1990, and in 1991 as well. In addition, it contains the world's greatest natural gas reserves, and one of the world's largest reserves of coal. Russian oil exports to Eastern Europe are transported primarily by pipeline through Ukraine, and to Western Europe via tanker through ports in the Black Sea and the Baltics. Besides the current natural gas production in West Siberia, vast amounts of natural gas are also believed to lie beneath the Arctic Ocean. In 1990, Russia produced over 10 million barrels per day of oil (91% of the total produced in the former Soviet Union), 23 trillion cubic feet of natural gas (79% of the Soviet total), and almost 400 million metric tons of coal (56% of the Soviet total). In the same year, the country consumed nearly 5 million barrels per day of oil, 17 trillion cubic feet of natural gas and virtually all of its coal production. Production of coal declined about 10% in 1991 while demand fell 22%. Total Russian crude oil-refining capacity is currently about 8 million barrels per day, accounting for around two-thirds of total Soviet refining capacity. These refineries are, in general, relatively antiquated and technically unsophisticated, with much less reforming and cracking capacity than US refineries. Production levels are difficult to maintain because of the lack of capital. Western countries are currently negotiating over a European Energy Charter which should help long-term energy development.

Russia has eleven operating nuclear reactors of the RBMK-Chernobyl type. A major effort to determine the best means of dealing with these reactors is under way with assistance of the IEA, the European Community and the World Bank.

A number of studies indicate that there would be a significant scope for improving energy efficiency and fuel-switching once the political and economic situation has stabilised and energy prices will have risen closer to world market levels.

Russian Federation

Key Energy and Environmental Data
(1990 data)

Adjusted TPES (Mtoe):	856.00
% Non-OECD TPES:	22.93
% Total World TPES:	11.02
Per Capita TPES (toe per person):	5.77
Per Capita TFC (toe per person):	3.33
TPES/GDP ratio (toe per US$1 000 1985):	N.A.
TFC/GDP ratio (toe per US$1 000 1985):	N.A.
Energy-related CO_2 emissions per capita (t CO_2 per person):	16.19
Energy-related CO_2 emissions per unit GDP (t CO_2 per US$1 000 1985):	N.A.
% Total energy-related Non-OECD CO_2 emissions:	21.49
% Total energy-related World CO_2 emissions:	11.13
% Population growth (yearly average growth 1981-1990):	N.A.

SAUDI ARABIA

1. APPARENT POSITION

Saudi Arabia is actively participating in the INC and the IPCC. It signed the June 1991 Beijing Declaration. It attended UNCED in Rio in June 1992 but did not sign the Convention. It has taken the lead in the IPCC's Working Group III's Subgroup on Energy and Industry in a study to assess the economic impact of climate change response measures on developing countries.

At the second session of the INC in Geneva in June 1991, Saudi Arabia, along with the Soviet Union and Kuwait, provided a paper which stated: "The specific difficulties of those countries, particularly developing countries, whose economies are highly dependent on fossil fuel production and exportation, as a consequence of action taken on limiting greenhouse gas emissions, should be taken into account".

2. FACTORS INFLUENCING DECISIONS

Saudi Arabia is one of the major oil producers and exporters and has over one-quarter of the world's proven crude oil reserves. The economy is highly dependent on oil and in 1989, oil exports represented 87% of total export revenue. Saudi Arabia is completely dependent on oil and natural gas for all its energy requirements.

Between 1973 and 1989, GDP grew at an annual rate of 2.7% while total primary energy supply grew at a rate of 15.4%, in large part for its growing energy-intensive industrial sector.

Saudi Arabia

Key Energy and Environmental Data
(1990 data)

Adjusted TPES (Mtoe):	73.86
% Non-OECD TPES:	1.98
% Total World TPES:	0.95
Per Capita TPES (toe per person):	4.97
Per Capita TFC (toe per person):	3.23
TPES/GDP ratio (toe per US$1 000 1985):	0.73
TFC/GDP ratio (toe per US$1 000 1985):	0.48
Energy-related CO_2 emissions per capita (t CO_2 per person):	13.59
Energy-related CO_2 emissions per unit GDP (t CO_2 per US$1 000 1985):	2.00
% Total energy-related Non-OECD CO_2 emissions:	1.81
% Total energy-related World CO_2 emissions:	0.94
% Population growth (yearly average growth 1981-1990):	5.73

SOUTH AFRICA

1. POSITION

The government is in the process of formulating policies on the issue of global climate change. To this end the Interdepartmental Co-ordinating Committee for Global Environmental Change was established in 1991 to "co-ordinate the action required for formulating broad national policy and strategy on global environmental change". There are six groups working on local policy and strategy implications. They cover: terrestrial environment (including agriculture, forestry and conservation); water resources and catchments; air quality; energy use; economics, trade technology and tourism; and the marine and coastal environment. The Department of Environmental Affairs is acting as the lead agent with other government departments participating. The task groups will involve a broad spectrum of experts from the public, private and research sectors. The policy formulation initiative is supported by an active programme of scientific research and environmental monitoring.

South Africa was not present at UNCED and had not signed the Convention at the time of publication.

2. FACTORS INFLUENCING DECISIONS

There are two main reasons for the government's interest in global climate change: first is because of the potential vulnerability of South Africa to such change. Much of South Africa is arid and the environment is subject to substantial natural variation. The implications of possible climate change for local industry, agriculture, forestry and marine resources and thus the economy are enormous. The second reason is that there is concern about the environment in general and a desire to play a responsible role regionally and globally in respect of global environmental change.

South Africa is the largest non-OECD hard coal exporter and ranks third in the world. South Africa also has the fifth largest hard coal reserves in the world. The country is highly dependent on fossil fuels. In 1989, 82% of TPES was coal and a further 9% was petroleum. In order to reduce reliance on imports, coal is used as a feedstock to produce synthetic oil. In the mid-1980s, it was the world leader in using synthetic fuel technology.

South Africa

Key Energy and Environmental Data
(1990 data):

Adjusted TPES (Mtoe):	99.48
% Non-OECD TPES:	2.66
% Total World TPES:	1.28
Per Capita TPES (toe per person):	2.82
Per Capita TFC (toe per person):	1.29
TPES/GDP ratio (toe per US$1 000 1985):	1.63
TFC/GDP ratio (toe per US$1 000 1985):	0.75
Energy-related CO_2 emissions per capita (t CO_2 per person):	9.52
Energy-related CO_2 emissions per unit GDP (t CO_2 per US$1 000 1985):	5.53
% Total energy-related Non-OECD CO_2 emissions:	3.01
% Total energy-related World CO_2 emissions:	1.56
% Population growth (yearly average growth 1981-1990):	2.45

UKRAINE

1. APPARENT POSITION

Ukraine attended all the sessions of the INC, even before the dissolution of the USSR, because it has traditionally had a separate seat in the United Nations. However, it was much less active than the USSR's delegation in the negotiations. Ukraine attended UNCED where it signed the Convention.

2. FACTORS INFLUENCING DECISIONS

On 1 December 1991 the people of Ukraine voted for independence from the USSR in a nationwide referendum. Ukraine has remained relatively free of domestic strife even after independence and despite the existence of large groups of different ethnic origins. Ukraine began radical economic reform in 1992 by drastically cutting price subsidies on most products and services. A package of reform measures, including privatisation and allowance for foreign investment, has begun to be introduced.

Ukraine possesses a well developed and diverse industrial base and is rich in natural resources. Ukraine plays a critical role in the energy picture of the former Soviet republics. First, Ukraine contains the Donets Basin, which is the largest coal-producing area in the former Soviet Union. Second, Ukraine is a major centre for heavy machinery and industrial equipment. Coal mining and metallurgy accounted for more than 40% of all industrial assets in 1990. Major upheavals are expected in these two areas. The coal industry is facing uncertainty because many of the seams are very deep and highly dangerous. The industry lacks modern machinery, so many of the mines could be abandoned. No subsidies are envisioned for the industry.

Ukraine is heavily dependent on Russia for oil and gas. At present Ukraine is only receiving about 30% of its former oil supplies from Russia which has led to a tenfold increase in the price of oil. However, although the country is a net energy importer, it serves a crucial strategic function as the major export route for Russian energy exports (mainly via pipeline) to Eastern and Western Europe. In fact, along with Belarus, Ukraine has a stranglehold on much of Russia's oil and gas exports to Eastern (and to some extent Western) Europe. Ukraine also contains a heavy concentration of nuclear power plans, with 15 of the 43 Soviet-designed nuclear generating units.

Overall, Ukraine consumes about 54.7 million tons of oil, 4.2 trillion cubic feet of natural gas, 135 million metric tons of coal (of which less than 10% is brown coal), and 256 billion kWh of electricity. It produces about 5 million tons of oil, 1 trillion cubic feet of natural gas, 183 million short tons of coal, and 278 billion kWh of electricity. Ukraine also possesses 1.1 million barrels per day of crude oil-refining capacity.

Ukraine

Key Energy and Environmental Data
(1990 data)

Adjusted TPES (Mtoe):	223.83
% Non-OECD TPES:	6.00
% Total World TPES:	2.88
Per Capita TPES (toe per person):	4.32
Per Capita TFC (toe per person):	N.A.
TPES/GDP ratio (toe per US$1 000 1985):	N.A.
TFC/GDP ratio (toe per US$1 000 1985):	N.A.
Energy-related CO_2 emissions per capita (t CO_2 per person):	12.70
Energy-related CO_2 emissions per unit GDP (t CO_2 per US$1 000 1985):	N.A.
% Total energy-related Non-OECD CO_2 emissions:	5.90
% Total energy-related World CO_2 emissions:	3.06
% Population growth (yearly average growth 1981-1990):	N.A.

BEIJING MINISTERIAL DECLARATION
ON ENVIRONMENT AND DEVELOPMENT

On 18-19 June 1991, 41 ministers from developing countries met in Beijing to discuss environment and development issues. Of the 41 Ministers, five came from the top 13 non-OECD carbon dioxide emitting countries: Brazil, China, India, Mexico and Saudi Arabia.

Some of the relevant passages related to climate change are:

- "environmental protection and sustainable development is a matter of common concern to humankind. . . we hereby reaffirm out solemn commitment to participating actively, on the basis of differentiated responsibility. . .".

- "the right to development of the developing countries must be fully recognised, and the adoption of measures for the protection of the global environment should support their economic growth and development".

- "Each country must be enabled to determine the pace of transition, based on the adaptive capacity of its economic, social and cultural ethos and capabilities".

- "The environmental problems of the developing countries arise from the conditions of poverty".

- "International co-operation . . . should be based on the principle of equality among sovereign states. The developing countries have the sovereign right to use their own natural resources in keeping with their developmental and environmental objectives and priorities".

- ". . . the developed countries bear the main responsibility for the degradation of the global environment. Ever since the Industrial Revolution, the developed countries have over-exploited the world's natural resources through unsustainable patterns of production and consumption, causing damage to the global environment, to the detriment of the developing countries".

- "The developed countries . . . must take the lead in eliminating the damage to the environment as well as in assisting the developing countries to deal with the problems facing them".

- "The developing countries need adequate, new and additional financial resources to be able to address effectively the environmental and developmental problems confronting them. There should be preferential and non-commercial transfer of environmentally sound technologies to the developing countries".

- "Responsibility for the emission of green-house gases should be viewed both in historical and cumulative terms, and in terms of current emissions. On the basis of the principle of equity, those developed countries which have contaminated more must contribute more. Developed countries should therefore commit themselves to adopting measures to half human-induced climate change and to setting up mechanisms to guarantee the environmental security and development of the developing countries. . . ".

- "The Convention must include, *inter alia,* firm commitments by developed countries towards the transfer of technology to developing countries, the establishment of a separate funding mechanism and the development of the economically viable new and renewable energy sources . . . In addition, the developing countries must be provided with the full scientific, technical and financial co-operation necessary to cope with the adverse impacts of climate change".

- "Issues of intellectual property rights must be satisfactorily resolved so that they do not become an obstacle to the transfer of technology. . .".

- "a special Green Fund should be established to provide adequate and additional financial assistance . . .".

- "The relevant agreements reached at the Conference [UNCED in Brazil, 1992] must provide guidance to international deliberations on trade, finance, technology and other similar issues. The interlinkages, where relevant, should be incorporated in each".

SECTION 4

UNITED NATIONS FRAMEWORK CONVENTION ON CLIMATE CHANGE

THE PARTIES TO THIS CONVENTION,

Acknowledging that change in the Earth's climate and its adverse effects are a common concern of humankind,

Concerned that human activities have been substantially increasing the atmospheric concentrations of greenhouse gases, that these increases enhance the natural greenhouse effect, and that this will result on average in an additional warming of the Earth's surface and atmosphere and may adversely affect natural ecosystems and humankind,

Noting that the largest share of historical and current global emissions of greenhouse gases has originated in developed countries, that per capita emissions in developing countries are still relatively low and that the share of global emissions originating in developing countries will grow to meet their social and development needs,

Aware of the role and importance in terrestrial and marine ecosystems of sinks and reservoirs of greenhouse gases,

Noting that there are many uncertainties in predictions of climate change, particularly with regard to the timing, magnitude and regional patterns thereof,

Acknowledging that the global nature of climate change calls for the widest possible co-operation by all countries and their participation in an effective and appropriate international response, in accordance with their common but differentiated responsibilities and respective capabilities and their social and economic conditions,

Recalling the pertinent provisions of the Declaration of the United Nations Conference on the Human Environment, adopted at Stockholm on 16 June 1972,

Recalling also that States have, in accordance with the Charter of the United Nations and the principles of international law, the sovereign right to exploit their own resources pursuant to their own environmental and developmental policies, and the responsibility to ensure that activities within their jurisdiction or control do not cause damage to the environment of other States or of areas beyond the limits of national jurisdiction,

Reaffirming the principle of sovereignty of States in international co-operation to address climate change,

Recognizing that States should enact effective environmental legislation, that environmental standards, management objectives and priorities should reflect the environmental and developmental context to which they apply, and that standards applied by some countries may be inappropriate and of unwarranted economic and social cost to other countries, in particular developing countries,

Recalling the provisions of General Assembly resolution 44/228 of 22 December 1989 on the UNCED, and resolutions 43/53 of 6 December 1988, 44/207 of 22 December 1989, 45/212 of 21 December 1990 and 46/169 of 19 December 1991 on protection of global climate for present and future generations of mankind,

Recalling also the provisions of General Assembly resolution 44/206 of 22 December 1989 on the possible adverse effects of sea level rise on islands and coastal areas, particularly low-lying coastal areas and the pertinent provisions of General Assembly resolution 44/172 of 19 December 1989 on the implementation of the Plan of Action to Combat Desertification,

Recalling further the Vienna Convention for the Protection of the Ozone Layer, 1985, and the Montreal Protocol on Substances that Deplete the Ozone Layer, 1987, as adjusted and amended on 29 June 1990,

Noting the Ministerial Declaration of the Second World Climate Conference adopted on 7 November 1990,

Conscious of the valuable analytical work being conducted by many States on climate change and of the important contributions of the World Meteorological Organization, the United Nations Environment Programme and other organs, organizations and bodies of the United Nations system, as well as other international and intergovernmental bodies, to the exchange of results of scientific research and the co-ordination of research,

Recognizing that steps required to understand and address climate change will be environmentally, socially and economically most effective if they are based on relevant scientific, technical and economic considerations and continually re-evaluated in the light of new findings in these areas,

Recognizing that various actions to address climate change can be justified economically in their own right and can also help in solving other environmental problems,

Recognizing also the need for developed countries to take immediate action in a flexible manner on the basis of clear priorities, as a first step towards comprehensive response strategies at the global, national and, where agreed, regional levels that take into account all greenhouse gases, with due consideration of their relative contributions to the enhancement of the greenhouse effect,

Recognizing further that low-lying and other small island countries, countries with low-lying coastal, arid and semi-arid areas or areas liable to floods, drought and desertification, and developing countries with fragile mountainous ecosystems are particularly vulnerable to the adverse effects of climate change,

Recognizing the special difficulties of those countries, especially developing countries, whose economies are particularly dependent on fossil fuel production, use and exportation, as a consequence of action taken on limiting greenhouse gas emissions,

Affirming that responses to climate change should be coordinated with social and economic development in an integrated manner with a view to avoiding adverse impacts on the latter, taking into full account the legitimate priority needs of developing countries for the achievement of sustained economic growth and the eradication of poverty,

Recognizing that all countries, especially developing countries, need access to resources required to achieve sustainable social and economic development and that, in order for developing countries to progress towards that goal, their energy consumption will need to grow taking into account the possibilities for achieving greater energy efficiency and for controlling greenhouse gas emissions in general, including through the application of new technologies on terms which make such an application economically and socially beneficial,

Determined to protect the climate system for present and future generations,

Have agreed as follows:

ARTICLE 1:   **Definitions**[1]

For the purposes of this Convention:

1. "Adverse effects of climate change" means changes in the physical environment or biota resulting from climate change which have significant deleterious effects on the composition, resilience or productivity of natural and managed ecosystems or on the operation of socio-economic systems or on human health and welfare.

---

1. Titles of articles are included solely to assist the reader.

2. "Climate change" means a change of climate which is attributed directly or indirectly to human activity that alters the composition of the global atmosphere and which is in addition to natural climate variability observed over comparable time periods.

3. "Climate system" means the totality of the atmosphere, hydrosphere, biosphere and geosphere and their interactions.

4. "Emissions" means the release of greenhouse gases and/or their precursors into the atmosphere over a specified area and period of time.

5. "Greenhouse gases" means those gaseous constituents of the atmosphere, both natural and anthropogenic, that absorb and re-emit infrared radiation.

6. "Regional economic integration organization" means an organization constituted by sovereign States of a given region which has competence in respect of matters governed by this Convention or its protocols and has been duly authorized, in accordance with its internal procedures, to sign, ratify, accept, approve or accede to the instruments concerned.

7. "Reservoir" means a component or components of the climate system where a greenhouse gas or a precursor of a greenhouse gas is stored.

8. "Sink" means any process, activity or mechanism which removes a greenhouse gas, an aerosol or a precursor of a greenhouse gas from the atmosphere.

9. "Source" means any process or activity which releases a greenhouse gas, an aerosol or a precursor of a greenhouse gas into the atmosphere.

ARTICLE 2:   **Objective**

The ultimate objective of this Convention and any related legal instruments that the Conference of the Parties may adopt is to achieve, in accordance with the relevant provisions of the Convention, stabilization of greenhouse gas concentrations in the atmosphere at a level that would prevent dangerous anthropogenic interference with the climate system. Such a level should be achieved within a time frame sufficient to allow ecosystems to adapt naturally to climate change, to ensure that food production is not threatened and to enable economic development to proceed in a sustainable manner.

ARTICLE 3:   **Principles**

In their actions to achieve the objective of the Convention and to implement its provisions, the Parties shall be guided, *inter alia,* by the following:

1. The Parties should protect the climate system for the benefit of present and future generations of humankind, on the basis of equity and in accordance with their common but differentiated responsibilities and respective capabilities. Accordingly, the developed country Parties should take the lead in combating climate change and the adverse effects thereof.

2. The specific needs and special circumstances of developing country Parties, especially those that are particularly vulnerable to the adverse effects of climate change, and of those Parties, especially developing country Parties, that would have to bear a disproportionate or abnormal burden under the Convention should be given full consideration.

3. The Parties should take precautionary measures to anticipate, prevent or minimize the causes of climate change and mitigate its adverse effects. Where there are threats of serious or irreversible damage, lack of full scientific certainty should not be used as a reason for postponing such measures, taking into account that policies and measures to deal with climate change should be cost-effective so as to ensure global benefits at the lowest possible cost. To achieve this, such policies and measures should take into account different socio-economic contexts, be comprehensive, cover all relevant sources, sinks and reservoirs of greenhouse gases and adaptation, and comprise all economic sectors. Efforts to address climate change may be carried out co-operatively by interested Parties.

4. The Parties have a right to, and should, promote sustainable development. Policies and measures to protect the climate system against human-induced change should be appropriate for the specific conditions of each Party and should be integrated with national development programmes, taking into account that economic development is essential for adopting measures to address climate change.

5. The Parties should co-operate to promote a supportive and open international economic system that would lead to sustainable economic growth and development in all Parties, particularly developing country Parties, thus enabling them better to address the problems of climate change. Measures taken to combat climate change, including unilateral ones, should not constitute a means of arbitrary or unjustifiable discrimination or a disguised restriction on international trade.

ARTICLE 4:   **Commitments**

1. All Parties, taking into account their common but differentiated responsibilities and their specific national and regional development priorities, objectives and circumstances, shall:

   (a) Develop, periodically update, publish and make available to the Conference of the Parties, in accordance with Article 12, national inventories of anthropogenic emissions by sources and removals by sinks of all greenhouse gases not controlled by the Montreal Protocol, using comparable methodologies to be agreed upon by the Conference of the Parties;

   (b) Formulate, implement, publish and regularly update national and, where appropriate, regional programmes containing measures to mitigate climate change by addressing anthropogenic emissions by sources and removals by sinks of all greenhouse gases not controlled by the Montreal Protocol, and measures to facilitate adequate adaptation to climate change;

   (c) Promote and co-operate in the development, application and diffusion, including transfer, of technologies, practices and processes that control, reduce or prevent anthropogenic emissions of greenhouse gases not controlled by the Montreal Protocol in all relevant sectors, including the energy, transport, industry, agriculture, forestry and waste management sectors;

(d) Promote sustainable management, and promote and co-operate in the conservation and enhancement, as appropriate, of sinks and reservoirs of all greenhouse gases not controlled by the Montreal Protocol, including biomass, forests and oceans as well as other terrestrial, coastal and marine ecosystems;

(e) Co-operate in preparing for adaptation to the impacts of climate change; develop and elaborate appropriate and integrated plans for coastal zone management, water resources and agriculture, and for the protection and rehabilitation of areas, particularly in Africa, affected by drought and desertification, as well as floods;

(f) Take climate change considerations into account, to the extent feasible, in their relevant social, economic and environmental policies and actions, and employ appropriate methods, for example impact assessments, formulated and determined nationally, with a view to minimizing adverse effects on the economy, on public health and on the quality of the environment, of projects or measures undertaken by them to mitigate or adapt to climate change;

(g) Promote and co-operate in scientific, technological, technical, socio-economic and other research, systematic observation and development of data archives related to the climate system and intended to further the understanding and to reduce or eliminate the remaining uncertainties regarding the causes, effects, magnitude and timing of climate change and the economic and social consequences of various response strategies;

(h) Promote and co-operate in the full, open and prompt exchange of relevant scientific, technological, technical, socio-economic and legal information related to the climate system and climate change, and to the economic and social consequences of various response strategies;

(i) Promote and co-operate in education, training and public awareness related to climate change and encourage the widest participation in this process, including that of non-governmental organizations; and

(j) Communicate to the Conference of the Parties information related to implementation, in accordance with Article 12.

2. The developed country Parties and other Parties included in Annex I commit themselves specifically as provided for in the following:

(a) Each of these Parties shall adopt national[1] policies and take corresponding measures on the mitigation of climate change, by limiting its anthropogenic emissions of greenhouse gases and protecting and enhancing its greenhouse gas sinks and reservoirs. These policies and measures will demonstrate that developed countries are taking the lead in modifying longer-term trends in anthropogenic emissions consistent with the objective of the Convention, recognizing that the return by the end of the present decade to earlier levels of anthropogenic emissions of carbon dioxide and other greenhouse gases not controlled by the Montreal Protocol would contribute to such modification, and taking into account the differences in these Parties' starting points and approaches, economic structures and resource bases, the need to maintain strong

---

1. This includes and measures adopted by regional economic integration organizations.

and sustainable economic growth, available technologies and other individual circumstances, as well as the need for equitable and appropriate contributions by each of these Parties to the global effort regarding that objective. These Parties may implement such policies and measures jointly with other Parties and may assist other Parties in contributing to the achievement of the objective of the Convention and, in particular, that of this subparagraph;

(b) In order to promote progress to this end, each of these Parties shall communicate, within six months of the entry into force of the Convention for it and periodically thereafter, and in accordance with Article 12, detailed information on its policies and measures referred to in subparagraph (a) above, as well as on its resulting projected anthropogenic emissions by sources and removals by sinks of greenhouse gases not controlled by the Montreal Protocol for the period referred to in subparagraph (a), with the aim of returning individually or jointly to their 1990 levels these anthropogenic emissions of carbon dioxide and other greenhouse gases not controlled by the Montreal Protocol. This information will be reviewed by the Conference of the Parties, at its first session and periodically thereafter, in accordance with Article 7;

(c) Calculations of emissions by sources and removals by sinks of greenhouse gases for the purposes of subparagraph (b) above should take into account the best available scientific knowledge, including of the effective capacity of sinks and the respective contributions of such gases to climate change. The Conference of the Parties shall consider and agree on methodologies for these calculations at its first session and review them regularly thereafter;

(d) The Conference of the Parties shall, at its first session, review the adequacy of subparagraphs (a) and (b) above. Such review shall be carried out in the light of the best available scientific information and assessment on climate change and its impacts, as well as relevant technical, social and economic information. Based on this review, the Conference of the Parties shall take appropriate action, which may include the adoption of amendments to the commitments in subparagraphs (a) and (b) above. The Conference of the Parties, at its first session, shall also take decisions regarding criteria for joint implementation as indicated in subparagraph (a) above. A second review of subparagraphs (a) and (b) shall take place not later than 31 December 1998, and thereafter at regular intervals determined by the Conference of the Parties, until the objective of the Convention is met;

(e) Each of these Parties shall:

(i) co-ordinate as appropriate with other such Parties, relevant economic and administrative instruments developed to achieve the objective of the Convention; and

(ii) identify and periodically review its own policies and practices which encourage activities that lead to greater levels of anthropogenic emissions of greenhouse gases not controlled by the Montreal Protocol than would otherwise occur;

(f) The Conference of the Parties shall review, not later than 31 December 1998, available information with a view to taking decisions regarding such amendments to the lists in Annexes I and II as may be appropriate, with the approval of the Party concerned;

(g) Any Party not included in Annex I may, in its instrument of ratification, acceptance, approval or accession, or at any time thereafter, notify the Depositary that it intends to be bound by subparagraphs (a) and (b) above. The Depositary shall inform the other signatories and Parties of any such notification.

3.   The developed country Parties and other developed Parties included in Annex II shall provide new and additional financial resources to meet the agreed full costs incurred by developing country Parties in complying with their obligations under Article 12 paragraph 1. They shall also provide such financial resources, including for the transfer of technology, needed by the developing country Parties to meet the agreed full incremental costs of implementing measures that are covered by paragraph 1 of this Article and that are agreed between a developing country Party and the international entity or entities referred to in Article 11, in accordance with that Article. The implementation of these commitments shall take into account the need for adequacy and predictability in the flow of funds and the importance of appropriate burden sharing among the developed country Parties.

4.   The developed country Parties and other developed Parties included in Annex II shall also assist the developing country Parties that are particularly vulnerable to the adverse effects of climate change in meeting costs of adaptation to those adverse effects.

5.   The developed country Parties and other developed Parties included in Annex II shall take all practicable steps to promote, facilitate and finance, as appropriate, the transfer of, or access to, environmentally sound technologies and know-how to other Parties, particularly developing country Parties, to enable them to implement the provisions of the Convention. In this process, the developed country Parties shall support the development and enhancement of endogenous capacities and technologies of developing country Parties. Other Parties and organizations in a position to do so may also assist in facilitating the transfer of such technologies.

6.   In the implementation of their commitments under paragraph 2 above, a certain degree of flexibility shall be allowed by the Conference of the Parties to the Parties included in Annex I undergoing the process of transition to a market economy, in order to enhance the ability of these Parties to address climate change, including with regard to the historical level of anthropogenic emissions of greenhouse gases not controlled by the Montreal Protocol chosen as a reference.

7.   The extent to which developing country Parties will effectively implement their commitments under the Convention will depend on the effective implementation by developed country Parties of their commitments under the Convention related to financial resources and transfer of technology and will take fully into account that economic and social development and poverty eradication are the first and overriding priorities of the developing country Parties.

8.   In the implementation of the commitments in this Article, the Parties shall give full consideration to what actions are necessary under the Convention, including actions related to funding, insurance and the transfer of technology, to meet the specific needs and concerns of developing country Parties arising from the adverse effects of climate change and/or the impact of the implementation of response measures, especially on:

(a) Small island countries;

(b) Countries with low-lying coastal areas;

(c) Countries with arid and semi-arid areas, forested areas and areas liable to forest decay;

(d) Countries with areas prone to natural disasters;

(e) Countries with areas liable to drought and desertification;

(f) Countries with areas of high urban atmospheric pollution;

(g) Countries with areas with fragile ecosystems including mountainous ecosystems:

(h) Countries whose economies are highly dependent on income generated from the production, processing and export, and/or on consumption of fossil fuels and associated energy-intensive products; and

(i) Land-locked and transit countries.

Further, the Conference of the Parties may take actions, as appropriate, with respect to this paragraph.

9. The Parties shall take full account of the specific needs and special situations of the least developed countries in their actions with regard to funding and transfer of technology.

10. The Parties shall, in accordance with Article 10, take into consideration in the implementation of the commitments of the Convention the situation of Parties, particularly developing country Parties, with economies that are vulnerable to the adverse effects of the implementation of measures to respond to climate change. This applies notably to Parties with economies that are highly dependent on income generated from the production, processing and export, and/or consumption of fossil fuels and associated energy-intensive products and/or the use of fossil fuels for which such Parties have serious difficulties in switching to alternatives.

ARTICLE 5:   **Research and Systematic Observation**

In carrying out their commitments under Article 4, paragraph 1(g), the Parties shall:

(a) Support and further develop, as appropriate, international and intergovernmental programmes and networks or organizations aimed at defining, conducting, assessing and financing research, data collection and systematic observation, taking into account the need to minimize duplication of effort;

(b) Support international and intergovernmental efforts to strengthen systematic observation and national scientific and technical research capacities and capabilities, particularly in developing countries, and to promote access to, and the exchange of, data and analyses thereof obtained from areas beyond national jurisdiction; and

(c) Take into account the particular concerns and needs of developing countries and cooperate in improving their endogenous capacities and capabilities to participate in the efforts referred to in subparagraphs (a) and (b) above.

ARTICLE 6:   **Education, Training and Public Awareness**

In carrying out their commitments under Article 4, paragraph 1(i), the Parties shall:

(a) Promote and facilitate at the national and, as appropriate, subregional and regional levels, and in accordance with national laws and regulations, and within their respective capacities:

   (i)   the development and implementation of educational and public awareness programmes on climate change and its effects;

   (ii) public access to information on climate change and its effects;

   (iii) public participation in addressing climate change and its effects and developing adequate responses; and

   (iv) training of scientific, technical and managerial personnel.

(b) Co-operate in and promote, at the international level, and, where appropriate, using existing bodies:

   (i)   the development and exchange of educational and public awareness material on climate change and its effects; and

   (ii) the development and implementation of education and training programmes, including the strengthening of national institutions and the exchange or secondment of personnel to train experts in this field, in particular for developing countries.

ARTICLE 7:   **Conference of the Parties**

1. A Conference of the Parties is hereby established.

2. The Conference of the Parties, as the supreme body of this Convention, shall keep under regular review the implementation of the Convention and any related legal instruments that the Conference of the Parties may adopt, and shall make, within its mandate, the decisions necessary to promote the effective implementation of the Convention. To this end, it shall:

(a) Periodically examine the obligations of the Parties and the institutional arrangements under the Convention, in the light of the objective of the Convention, the experience gained in its implementation and the evolution of scientific and technological knowledge;

(b) Promote and facilitate the exchange of information on measures adopted by the Parties to address climate change and its effects, taking into account the differing circumstances, responsibilities and capabilities of the Parties and their respective commitments under the Convention;

(c) Facilitate, at the request of two or more Parties, the co-ordination of measures adopted by them to address climate change and its effects, taking into account the differing circumstances, responsibilities and capabilities of the Parties and their respective commitments under the Convention;

(d) Promote and guide, in accordance with the objective and provisions of the Convention, the development and periodic refinement of comparable methodologies, to be agreed on by the Conference of the Parties, inter alia, for preparing inventories of greenhouse gas emissions by sources and removals by sinks, and for evaluating the effectiveness of measures to limit the emissions and enhance the removals of these gases;

(e) Assess, on the basis of all information made available to it in accordance with the provisions of the Convention, the implementation of the Convention by the Parties, the overall effects of the measures taken pursuant to the Convention, in particular environmental, economic and social effects as well as their cumulative impacts and the extent to which progress towards the objective of the Convention is being achieved;

(f) Consider and adopt regular reports on the implementation of the Convention and ensure their publication;

(g) Make recommendations on any matters necessary for the implementation of the Convention;

(h) Seek to mobilize financial resources in accordance with Article 4, paragraphs 3, 4 and 5, and Article 11;

(i) Establish such subsidiary bodies as are deemed necessary for the implementation of the Convention;

(j) Review reports submitted by its subsidiary bodies and provide guidance to them;

(k) Agree upon and adopt, by consensus, rules of procedure and financial rules for itself and for any subsidiary bodies;

(l) Seek and utilize, where appropriate, the services and co-operation of, and information provided by, competent international organizations and intergovernmental and non-governmental bodies; and

(m) Exercise such other functions as are required for the achievement of the objective of the Convention as well as all other functions assigned to it under the Convention.

3. The Conference of the Parties shall, at its first session, adopt its own rules of procedure as well as those of the subsidiary bodies established by the Convention, which shall include decision-making procedures for matters not already covered by decision-making procedures stipulated in the Convention. Such procedures may include specified majorities required for the adoption of particular decisions.

4. The first session of the Conference of the Parties shall be convened by the interim secretariat referred to in Article 21 and shall take place not later than one year after the date of entry into force of the Convention. Thereafter, ordinary sessions of the Conference of the Parties shall be held every year unless otherwise decided by the Conference of the Parties.

5. Extraordinary sessions of the Conference of the Parties shall be held at such other times as may be deemed necessary by the Conference, or at the written request of any Party, provided that, within six months of the request being communicated to the Parties by the secretariat, it is supported by at least one-third of the Parties.

6. The United Nations, its specialized agencies and the International Atomic Energy Agency, as well as any State member thereof or observers thereto not Party to the Convention, may be represented at sessions of the Conference of the Parties as observers. Any body or agency, whether national or international, governmental or non-governmental, which is qualified in matters covered by the Convention, and which has informed the secretariat of its wish to be represented at a session of the Conference of the Parties as an observer, may be so admitted unless at least one-third of the Parties present object. The admission and participation of observers shall be subject to the rules of procedure adopted by the Conference of the Parties.

ARTICLE 8:   **Secretariat**

1. A secretariat is hereby established.
2. The functions of the secretariat shall be:
    (a) To make arrangements for sessions of the Conference of the Parties and its subsidiary bodies established under the Convention and to provide them with services as required;
    (b) To compile and transmit reports submitted to it;
    (c) To facilitate assistance to the Parties particularly developing country Parties, on request, in the compilation and communication of information required in accordance with the provisions of the Convention;
    (d) To prepare reports on its activities and present them to the Conference of the Parties;
    (e) To ensure the necessary co-ordination with the secretariats of other relevant international bodies;
    (f) To enter, under the overall guidance of the Conference of the Parties, into such administrative and contractual arrangements as may be required for the effective discharge of its functions; and
    (g) to perform the other secretariat functions specified in the Convention and in any of its protocols and such other functions as may be determined by the Conference of the Parties.
3. The Conference of the Parties, at its first session, shall designate a permanent secretariat and make arrangements for its functioning.

ARTICLE 9:   **Subsidiary Body for Scientific and Technological Advice**

1. A subsidiary body for scientific and technological advice is hereby established to provide the Conference of the Parties and, as appropriate, its other subsidiary bodies with timely information and advice on scientific and technological matters relating to the Convention. This body shall be open to participation by all Parties and shall be multidisciplinary. It shall comprise government representatives competent in the relevant field of expertise. It shall report regularly to the Conference of the Parties on all aspects of its work.

2. Under the guidance of the Conference of the Parties, and drawing upon existing competent international bodies, this body shall:

(a) Provide assessments of the state of scientific knowledge relating to climate change and its effects;

(b) Prepare scientific assessments on the effects of measures taken in the implementation of the Convention;

(c) Identify innovative, efficient and state-of-the-art technologies and know-how and advise on the ways and means of promoting development and/or transferring such technologies;

(d) provide advice on scientific programmes, international co-operation in research and development related to climate change, as well as on ways and means of supporting endogenous capacity-building in developing countries; and

(e) Respond to scientific, technological and methodological questions that the Conference of the Parties and its subsidiary bodies may put to the body.

3. The functions and terms of reference of this body may be further elaborated by the Conference of the Parties.

ARTICLE 10:   **Subsidiary Body for Implementation**

1. A subsidiary body for implementation is hereby established to assist the Conference of the Parties in the assessment and review of the effective implementation of the Convention. This body shall be open to participation by all Parties and comprise government representatives who are experts on matters related to climate change. It shall report regularly to the Conference of the Parties on all aspects of its work.

2. Under the guidance of the Conference of the Parties, this body shall:

(a) Consider the information communicated in accordance with Article 12, paragraph 1, to assess the overall aggregated effect of the steps taken by the Parties in the light of the latest scientific assessments concerning climate change;

(b) Consider the information communicated in accordance with Article 12, paragraph 2, in order to assist the Conference of the Parties in carrying out the reviews required by Article 4, paragraph 2(d); and

(c) Assist the Conference of the Parties, as appropriate, in the preparation and implementation of its decisions.

ARTICLE 11:   **Financial Mechanism**

1. A mechanism for the provision of financial resources on a grant or confessional basis, including for the transfer of technology, is hereby defined. It shall function under the guidance of and be accountable to the Conference of the Parties, which shall decide on its policies, programme priorities and eligibility criteria related to this Convention. lts operation shall be entrusted to one or more existing international entities.

2. The financial mechanism shall have an equitable and balanced representation of all Parties within a transparent system of governance.

3. The Conference of the Parties and the entity or entities entrusted with the operation of the financial mechanism shall agree upon arrangements to give effect to the above paragraphs, which shall include the following:

(a) Modalities to ensure that the funded projects to address climate change are in conformity with the policies, programme priorities and eligibility criteria established by the Conference of the Parties;

(b) Modalities by which a particular funding decision may be reconsidered in light of these policies, programme priorities and eligibility criteria;

(c) Provision by the entity or entities of regular reports to the Conference of the Parties on its funding operations, which is consistent with the requirement for accountability set out in paragraph 1 above; and

(d) Determination in a predictable and identifiable manner of the amount of funding necessary and available for the implementation of this Convention and the conditions under which that amount shall be periodically reviewed.

4. The Conference of the Parties shall make arrangements to implement the above-mentioned provisions at its first session, reviewing and taking into account the interim arrangements referred to in Article 21, paragraph 3, and shall decide whether these interim arrangements shall be maintained. Within four years thereafter, the Conference of the Parties shall review the financial mechanism and take appropriate measures.

5. The developed country Parties may also provide and developing country Parties avail themselves of, financial resources related to the implementation of the Convention through bilateral, regional and other multilateral channels.

ARTICLE 12:   **Communication of Information Related to Implementation**

1. In accordance with Article 4, paragraph 1, each Party shall communicate to the Conference of the Parties, through the secretariat, the following elements of information:

(a) A national inventory of anthropogenic emissions by sources and removals by sinks of all greenhouse gases not controlled by the Montreal Protocol, to the extent its capacities permit, using comparable methodologies to be promoted and agreed upon by the Conference of the Parties;

(b) A general description of steps taken or envisaged by the Party to implement the Convention; and

(c) Any other information that the Party considers relevant to the achievement of the objective of the Convention and suitable for inclusion in its communication, including, if feasible, material relevant for calculations of global emission trends.

2. Each developed country Party and each other Party included in Annex I shall incorporate in its communication the following elements of information:

(a) A detailed description of the policies and measures that it has adopted to implement its commitment under Article 4, paragraphs 2(a) and 2(b); and

(b) A specific estimate of the effects that the policies and measures referred to in subparagraph (a) immediately above will have on anthropogenic emissions by its sources and removals by its sinks of greenhouse gases during the period referred to in Article 4, paragraph 2(a).

3. In addition, each developed country Party and each other developed Party included in Annex II shall incorporate details of measures taken in accordance with Article 4, paragraphs 3, 4 and 5.

4. Developing country Parties may, on a voluntary basis, propose projects for financing, including specific technologies, materials, equipment, techniques or practices that would be needed to implement such projects, along with, if possible, an estimate of all incremental costs, of the reductions of emissions and increments of removals of greenhouse gases, as well as an estimate of the consequent benefits.

5. Each developed country Party and each other Party included in Annex I shall make its initial communication within six months of the entry into force of the Convention for that Party. Each Party not so listed shall make its initial communication within three years of the entry into force of the Convention for that Party, or of the availability of financial resources in accordance with Article 4, paragraph 3. Parties that are least developed countries may make their initial communication at their discretion. The frequency of subsequent communications by all Parties shall be determined by the Conference of the Parties, taking into account the differentiated timetable set by this paragraph.

6. Information communicated by Parties under this Article shall be transmitted by the secretariat as soon as possible to the Conference of the Parties and to any subsidiary bodies concerned. If necessary, the procedures for the communication of information may be further considered by the Conference of the Parties.

7. From its first session, the Conference of the Parties shall arrange for the provision to developing country Parties of technical and financial support, on request, in compiling and communicating information under this Article, as well as in identifying the technical and financial needs associated with proposed projects and response measures under Article 4. Such support may be provided by other Parties, by competent international organizations and by the secretariat, as appropriate.

8. Any group of Parties may, subject to guidelines adopted by the Conference of the Parties, and to prior notification to the Conference of the Parties, make a joint communication in fulfilment of their obligations under this Article, provided that such a communication includes information on the fulfilment by each of these Parties of its individual obligations under the Convention.

9. Information received by the secretariat that is designated by a Party as confidential, in accordance with criteria to be established by the Conference of the Parties, shall be aggregated by the secretariat to protect its confidentiality before being made available to any of the bodies involved in the communication and review of information.

10. Subject to paragraph 9 above, and without prejudice to the ability of any Party to make public its communication at any time, the secretariat shall make communications by Parties under this Article publicly available at the time they are submitted to the Conference of the Parties.

ARTICLE 13:   **Resolution of Questions Regarding Implementation**

The Conference of the Parties shall, at its first session, consider the establishment of a multilateral consultative process, available to Parties on their request, for the resolution of questions regarding the implementation of the Convention.

ARTICLE 14:   **Settlement of Disputes**

1.   In the event of a dispute between any two or more Parties concerning the interpretation or application of the Convention, the Parties concerned shall seek a settlement of the dispute through negotiation or any other peaceful means of their own choice.

2.   When ratifying, accepting, approving or acceding to the Convention, or at any time thereafter, a Party which is not a regional economic integration organization may declare in a written instrument submitted to the Depositary that, in respect of any dispute concerning the interpretation or application of the Convention, it recognizes as compulsory ipso facto and without special agreement, in relation to any Party accepting the same obligation:

(a) Submission of the dispute to the international Court of Justice, and/or

(b) Arbitration in accordance with procedures to be adopted by the Conference of the Parties as soon as practicable, in an annex on arbitration.

A Party which is a regional economic integration organization may make a declaration with like effect in relation to arbitration in accordance with the procedures referred to in sub-paragraph (b) above.

3.   A declaration made under paragraph 2 above shall remain in force until it expires in accordance with its terms or until three months after written notice of its revocation has been deposited with the Depositary.

4.   A new declaration, a notice of revocation or the expiry of a declaration shall not in any way affect proceedings pending before the International Court of Justice or the arbitral tribunal, unless the parties to the dispute otherwise agree.

5.   Subject to the operation of paragraph 2 above, if after twelve months following notification by one Party to another that a dispute exists between them, the Parties concerned have not been able to settle their dispute through the means mentioned in paragraph 1 above, the dispute shall be submitted, at the request of any of the parties to the dispute, to conciliation.

6.   A conciliation commission shall be created upon the request of one of the parties to the dispute. The commission shall be composed of an equal number of members appointed by each party concerned and a chairman chosen jointly by the members appointed by each party. The commission shall render a recommendatory award, which the parties shall consider in good faith.

7.   Additional procedures relating to conciliation shall be adopted by the Conference of the Parties, as soon as practicable, in an annex on conciliation.

8. The provisions of this Article shall apply to any related legal instrument which the Conference of the Parties may adopt, unless the instrument provides otherwise.

ARTICLE 15:   **Amendments to the Convention**

1. Any Party may propose amendments to the Convention.

2. Amendments to the Convention shall be adopted at an ordinary session of the Conference of the Parties. The text of any proposed amendment to the Convention shall be communicated to the Parties by the secretariat at least six months before the meeting at which it is proposed for adoption. The secretariat shall also communicate proposed amendments to the signatories to the Convention and, for information, to the Depositary.

3. The Parties shall make every effort to reach agreement on any proposed amendment to the Convention by consensus. If all efforts at consensus have been exhausted, and no agreement reached, the amendment shall as a last resort be adopted by a three-fourths majority vote of the Parties present and voting at the meeting. The adopted amendment shall be communicated by the secretariat to the Depositary, who shall circulate it to all Parties for their acceptance.

4. Instruments of acceptance in respect of an amendment shall be deposited with the Depositary. An amendment adopted in accordance with paragraph 3 above shall enter into force for those Parties having accepted it on the ninetieth day after the date of receipt by the Depositary of an instrument of acceptance by at least three-fourths of the Parties to the Convention.

5. The amendment shall enter into force for any other Party on the ninetieth day after the date on which that Party deposits with the Depositary its instrument of acceptance of the said amendment.

6. For the purposes of this Article, "Parties present and voting" means Parties present and casting an affirmative or negative vote.

ARTICLE 16:   **Adoption and amendment of annexes to the Convention**

1. Annexes to the Convention shall form an integral part thereof and, unless otherwise expressly provided, a reference to the Convention constitutes at the same time a reference to any annexes thereto. Without prejudice to the provisions of Article 14, paragraphs 2(b) and 7, such annexes shall be restricted to lists, forms and any other material of a descriptive nature that is of a scientific, technical, procedural or administrative character.

2. Annexes to the Convention shall be proposed and adopted in accordance with the procedure set forth in Article 15, paragraphs 2, 3, and 4.

3. An annex that has been adopted in accordance with paragraph 2 above shall enter into force for all Parties to the Convention six months after the date of the communication by the Depositary to such Parties of the adoption of the annex, except for those Parties that have notified the Depositary, in writing, within that period of their non-acceptance of the annex.

The annex shall enter into force for Parties which withdraw their notification of non-acceptance on the ninetieth day after the date on which withdrawal of such notification has been received by the Depositary.

4. The proposal, adoption and entry into force of amendments to annexes to the Convention shall be subject to the same procedure as that for the proposal, adoption and entry into force of annexes to the Convention in accordance with paragraphs 2 and 3 above.

5. If the adoption of an annex or an amendment to an annex involves an amendment to the Convention, that annex or amendment to an annex shall not enter into force until such time as the amendment to the Convention enters into force.

ARTICLE 17:   **Protocols**

1. The Conference of the Parties may, at any ordinary session, adopt protocols to the Convention.

2. The text of any proposed protocol shall be communicated to the Parties by the secretariat at least six months before such a session.

3. The requirements for the entry into force of any protocol shall be established by that instrument.

4. Only Parties to the Convention may be Parties to a protocol.

5. Decisions under any protocol shall be taken only by the Parties to the protocol concerned.

ARTICLE 18:   **Right to Vote**

1. Each Party to the Convention shall have one vote, except as provided for in paragraph 2 below.

2. Regional economic integration organizations, in matters within their competence, shall exercise their right to vote with a number of votes equal to the number of their member States that are Parties to the Convention. Such an organization shall not exercise its right to vote if any of its member States exercises its right, and vice versa.

ARTICLE 19:   **Depositary**

The Secretary-General of the United Nations shall be the Depositary of the Convention and of protocols adopted in accordance with Article 17.

ARTICLE 20:   **Signature**

This Convention shall be open for signature by States Members of the United Nations or of any of its specialized agencies or that are Parties to the Statute of the International Court of Justice and by regional economic integration organizations at Rio de Janeiro, during the UNCED, and thereafter at United Nations Headquarters in New York from 20 June 1992 to 19 June 1993.

ARTICLE 21:   **Interim Arrangements**

1. The secretariat functions referred to in Article 8 will be carried out on an interim basis by the secretariat established by the General Assembly of the United Nations in its resolution 45/212 of 21 December 1990, until the completion of the first session of the Conference of the Parties.

2. The head of the interim secretariat referred to in paragraph 1 above will co-operate closely with the Intergovernmental Panel on Climate Change to ensure that the Panel can respond to the need for objective scientific and technical advice. Other relevant scientific bodies could also be consulted.

3. The Global Environment Facility of the United Nations Development Programme, the United Nations Environment Programme and the International Bank for Reconstruction and Development shall be the international entity entrusted with the operation of the financial mechanism referred to in Article 11 on an interim basis. In this connection, the Global Environment Facility should be appropriately restructured and its membership made universal to enable it to fulfil the requirements of Article 11.

ARTICLE 22:   **Ratification, Acceptance, Approval or Accession**

1. The Convention shall be subject to ratification, acceptance, approval or accession by States and by regional economic integration organizations. It shall be open for accession from the day after the date on which the Convention is closed for signature. Instruments of ratification, acceptance, approval or accession shall be deposited with the Depositary.

2. Any regional economic integration organization which becomes a Party to the Convention without any of its member States being a Party shall be bound by all the obligations under the Convention. In the case of such organizations, one or more of whose member States is a Party to the Convention, the organization and its member States shall decide on their respective responsibilities for the performance of their obligations under the Convention. In such cases, the organization and the member States shall not be entitled to exercise rights under the Convention concurrently.

3. In their instruments of ratification, acceptance, approval or accession, regional economic integration organizations shall declare the extent of their competence with respect to the matters governed by the Convention. These organizations shall also inform the Depositary, who shall in turn inform the Parties, of any substantial modification in the extent of their competence.

ARTICLE 23:   **Entry into Force**

1. The Convention shall enter into force on the ninetieth day after the date of deposit of the fiftieth instrument of ratification, acceptance, approval or accession.

2. For each State or regional economic integration organization that ratifies, accepts or approves the Convention or accedes thereto after the deposit of the fiftieth instrument of ratification, acceptance, approval or accession, the Convention shall enter into force on the

ninetieth day after the date of deposit by such State or regional economic integration organization of its instrument of ratification, acceptance, approval or accession.

3. For the purposes of paragraphs 1 and 2 above, any instrument deposited by a regional economic integration organization shall not be counted as additional to those deposited by States members of the organization.

ARTICLE 24:   **Reservations**

No reservations may be made to the Convention.

ARTICLE 25:   **Withdrawal**

1. At any time after three years from the date on which the Convention has entered into force for a Party, that Party may withdraw from the Convention by giving written notification to the Depositary.
2. Any such withdrawal shall take effect upon expiry of one year from the date of receipt by the Depositary of the notification of withdrawal, or on such later date as may be specified in the notification of withdrawal.
3. Any Party that withdraws from the Convention shall be considered as also having withdrawn from any protocol to which it is a Party.

ARTICLE 26:   **Authentic Texts**

The original of this Convention, of which the Arabic, Chinese, English, French, Russian and Spanish texts are equally authentic, shall be deposited with the Secretary-General of the United Nations.

In Witness where of the undersigned, being duly authorized to that effect, have signed this Convention.

Done at New York this ninth day of May one thousand nine hundred and ninety-two.

ANNEX I

Australia
Austria
Belarus
Belgium
Bulgaria [1]
Canada
Czechoslovakia [1]
Denmark
European Community
Estonia [1]
Finland
France
Germany
Greece
Hungary [1]
Iceland
Ireland
Italy
Japan
Latvia [1]
Lithuania [1]
Luxembourg
Netherlands
New Zealand
Norway
Poland [1]
Portugal

---

1. Countries that are undergoing the process of transition to a market economy.

Romania[1]
Russian Federation[1]
Spain
Sweden
Switzerland
Turkey
Ukraine[1]
United Kingdom of Great Britain and Northern Ireland
United States of America

---

1. Countries that are undergoing the process of transition to a market economy.

ANNEX II

Australia
Austria
Belgium
Canada
Denmark
European Community
Finland
France
Germany
Greece
Iceland
Ireland
Italy
Japan
Luxembourg
Netherlands
New Zealand
Norway
Portugal
Spain
Sweden
Switzerland
Turkey
United Kingdom of Great Britain and Northern Ireland
United States of America

RESOLUTION ADOPTED BY THE INTERGOVERNMENTAL NEGOTIATING COMMITTEE FOR A FRAMEWORK CONVENTION ON CLIMATE CHANGE

INC/1992/1. Interim Arrangements

The Intergovernmental Negotiating Committee for a Framework Convention on Climate Change,

Having agreed upon and adopted the text of the United Nations Framework Convention on Climate Change,

Considering that preparations are required for an early and effective operation of the Convention once it has entered into force,

Further considering that, in the interim arrangements, involvement in the negotiations of all participants in the Committee is essential,

Recalling General Assembly resolutions 45/212 of 21 December 1990 and 46/169 of 19 December 1991,

1. **Calls upon** all States and regional economic integration organizations entitled to do so to sign the Convention during the UNCED in Rio de Janeiro or at the earliest subsequent opportunity and thereafter to ratify, accept, approve or accede to the Convention;

2. **Requests** the Secretary-General to make the necessary arrangements for convening a session of the Committee, in accordance with paragraph 4 of General Assembly resolution 46/169, to prepare for the first session of the Conference of the Parties as specified in the Convention;

3. **Requests further** the Secretary-General to make recommendations to the General Assembly at its forty-seventh session regarding arrangements for further sessions of the Committee until the entry into force of the Convention;

4. **Invites** the Secretary-General to include in his report to the General Assembly, as required in paragraphs 4 and 9 of resolution 46/169, proposals that would enable the Secretariat established under resolution 45/212 to continue its activities until the designation of the secretariat of the Convention by the Conference of the Parties;

5. **Appeals** to Governments and organizations to make voluntary contributions to the extrabudgetary funds established under General Assembly resolution 45/212 in order to contribute to the costs of the interim arrangements, and to ensure full and effective participation of developing countries, in particular the least developed countries and small island developing countries, as well as developing countries stricken by drought and desertification, in all the sessions of the Committee;

6. **Invites** States and regional economic integration organizations entitled to sign the Convention to communicate as soon as feasible to the head of the secretariat information regarding measures consistent with the provisions of the Convention pending its entry into force.

9 May 1992

GLOSSARY

CFC:	Chlorofluorocarbons
CH_4:	Methane
Convention:	Framework Convention on Climate Change
CO_2:	Carbon dioxide
EC:	European Community
EFTA:	European Free Trade Association
GDP:	Gross Domestic Product
GHG:	Greenhouse gases
GWP:	Global warming potential
IEA:	International Energy Agency
INC:	Intergovernmental Negotiating Committee on a Framework Convention for Climate Change
IPCC:	Intergovernmental Panel on Climate Change
ktons:	Thousand tons
Mtoe:	Million tons of oil equivalent
Mtons:	Million tons
NMP:	Non-Montreal Protocol
NO_x:	Oxides of nitrogen
OECD:	Organisation for Economic Co-operation and Development
SO_2:	Sulphur dioxide
t:	Tons
TFC:	Total Final Consumption
TPES:	Total Primary Energy Supply
UNCED:	United Nations Conference on Environment and Development

MAIN SALES OUTLETS OF OECD PUBLICATIONS
PRINCIPAUX POINTS DE VENTE DES PUBLICATIONS DE L'OCDE

ARGENTINA – ARGENTINE
Carlos Hirsch S.R.L.
Galería Güemes, Florida 165, 4° Piso
1333 Buenos Aires   Tel. (1) 331.1787 y 331.2391
Telefax: (1) 331.1787

AUSTRALIA – AUSTRALIE
D.A. Book (Aust.) Pty. Ltd.
648 Whitehorse Road, P.O.B 163
Mitcham, Victoria 3132   Tel. (03) 873.4411
Telefax: (03) 873.5679

AUSTRIA – AUTRICHE
Gerold & Co.
Graben 31
Wien I   Tel. (0222) 533.50.14

BELGIUM – BELGIQUE
Jean De Lannoy
Avenue du Roi 202
B-1060 Bruxelles   Tel. (02) 538.51.69/538.08.41
Telefax: (02) 538.08.41

CANADA
Renouf Publishing Company Ltd.
1294 Algoma Road
Ottawa, ON K1B 3W8   Tel. (613) 741.4333
Telefax: (613) 741.5439
Stores:
61 Sparks Street
Ottawa, ON K1P 5R1   Tel. (613) 238.8985
211 Yonge Street
Toronto, ON M5B 1M4   Tel. (416) 363.3171
Les Éditions La Liberté Inc.
3020 Chemin Sainte-Foy
Sainte-Foy, PQ G1X 3V6   Tel. (418) 658.3763
Telefax: (418) 658.3763

Federal Publications
165 University Avenue
Toronto, ON M5H 3B8   Tel. (416) 581.1552
Telefax: (416) 581.1743

CHINA – CHINE
China National Publications Import
Export Corporation (CNPIEC)
P.O. Box 88
Beijing   Tel. 403.5533
Telefax: 401.5664

DENMARK – DANEMARK
Munksgaard Export and Subscription Service
35, Nørre Søgade, P.O. Box 2148
DK-1016 København K   Tel. (33) 12.85.70
Telefax: (33) 12.93.87

FINLAND – FINLANDE
Akateeminen Kirjakauppa
Keskuskatu 1, P.O. Box 128
00100 Helsinki   Tel. (358 0) 12141
Telefax: (358 0) 121.4441

FRANCE
OECD/OCDE
Mail Orders/Commandes par correspondance:
2, rue André-Pascal
75775 Paris Cedex 16   Tel. (33-1) 45.24.82.00
Telefax: (33-1) 45.24.85.00   or (33-1) 45.24.81.76
Telex: 620 160 OCDE
OECD Bookshop/Librairie de l'OCDE :
33, rue Octave-Feuillet
75016 Paris   Tel. (33-1) 45.24.81.67
(33-1) 45.24.81.81

Documentation Française
29, quai Voltaire
75007 Paris   Tel. 40.15.70.00
Gibert Jeune (Droit-Économie)
6, place Saint-Michel
75006 Paris   Tel. 43.25.91.19

Librairie du Commerce International
10, avenue d'Iéna
75016 Paris   Tel. 40.73.34.60
Librairie Dunod
Université Paris-Dauphine
Place du Maréchal de Lattre de Tassigny
75016 Paris   Tel. 47.27.18.56
Librairie Lavoisier
11, rue Lavoisier
75008 Paris   Tel. 42.65.39.95
Librairie L.G.D.J. - Montchrestien
20, rue Soufflot
75005 Paris   Tel. 46.33.89.85
Librairie des Sciences Politiques
30, rue Saint-Guillaume
75007 Paris   Tel. 45.48.36.02
P.U.F.
49, boulevard Saint-Michel
75005 Paris   Tel. 43.25.83.40
Librairie de l'Université
12a, rue Nazareth
13100 Aix-en-Provence   Tel. (16) 42.26.18.08
Documentation Française
165, rue Garibaldi
69003 Lyon   Tel. (16) 78.63.32.23

GERMANY – ALLEMAGNE
OECD Publications and Information Centre
Schedestrasse 7
D-W 5300 Bonn 1   Tel. (0228) 21.60.45
Telefax: (0228) 26.11.04

GREECE – GRÈCE
Librairie Kauffmann
Mavrokordatou 9
106 78 Athens   Tel. 322.21.60
Telefax: 363.39.67

HONG-KONG
Swindon Book Co. Ltd.
13–15 Lock Road
Kowloon, Hong Kong   Tel. 366.80.31
Telefax: 739.49.7ʳ

ICELAND – ISLANDE
Mál Mog Menning
Laugavegi 18, Pósthólf 392
121 Reykjavik   Tel. 162.35.23

INDIA – INDE
Oxford Book and Stationery Co.
Scindia House
New Delhi 110001   Tel.(11) 331.5896/5308
Telefax: (11) 332.5993
17 Park Street
Calcutta 700016   Tel. 240832

INDONESIA – INDONÉSIE
Pdii-Lipi
P.O. Box 4298
Jakarta 12042   Tel. 583467
Telex: 62 875

IRELAND – IRLANDE
TDC Publishers – Library Suppliers
12 North Frederick Street
Dublin 1   Tel. 74.48.35/74.96.77
Telefax: 74.84.16

ISRAEL
Electronic Publications only
Publications électroniques seulement
Sophist Systems Ltd.
71 Allenby Street
Tel-Aviv 65134   Tel. 3-29.00.21
Telefax: 3-29.92.39

ITALY – ITALIE
Libreria Commissionaria Sansoni
Via Duca di Calabria 1/1
50125 Firenze   Tel. (055) 64.54.15
Telefax: (055) 64.12.57
Via Bartolini 29
20155 Milano   Tel. (02) 36.50.83
Editrice e Libreria Herder
Piazza Montecitorio 120
00186 Roma   Tel. 679.46.28
Telefax: 678.47.51
Libreria Hoepli
Via Hoepli 5
20121 Milano   Tel. (02) 86.54.46
Telefax: (02) 805.28.86
Libreria Scientifica
Dott. Lucio de Biasio 'Aeiou'
Via Coronelli, 6
20146 Milano   Tel. (02) 48.95.45.52
Telefax: (02) 48.95.45.48

JAPAN – JAPON
OECD Publications and Information Centre
Landic Akasaka Building
2-3-4 Akasaka, Minato-ku
Tokyo 107   Tel. (81.3) 3586.2016
Telefax: (81.3) 3584.7929

KOREA – CORÉE
Kyobo Book Centre Co. Ltd.
P.O. Box 1658, Kwang Hwa Moon
Seoul   Tel. 730.78.91
Telefax: 735.00.30

MALAYSIA – MALAISIE
Co-operative Bookshop Ltd.
University of Malaya
P.O. Box 1127, Jalan Pantai Baru
59700 Kuala Lumpur
Malaysia   Tel. 756.5000/756.5425
Telefax: 755.4424

NETHERLANDS – PAYS-BAS
SDU Uitgeverij
Christoffel Plantijnstraat 2
Postbus 20014
2500 EA's-Gravenhage   Tel. (070 3) 78.99.11
Voor bestellingen: Tel. (070 3) 78.98.80
Telefax: (070 3) 47.63.51

NEW ZEALAND
NOUVELLE-ZÉLANDE
Legislation Services
P.O. Box 12418
Thorndon, Wellington   Tel. (04) 496.5652
Telefax: (04) 496.5698

NORWAY – NORVÈGE
Narvesen Info Center – NIC
Bertrand Narvesens vei 2
P.O. Box 6125 Etterstad
0602 Oslo 6   Tel. (02) 57.33.00
Telefax: (02) 68.19.01

PAKISTAN
Mirza Book Agency
65 Shahrah Quaid-E-Azam
Lahore 3   Tel. 66.839
Telex: 44886 UBL PK. Attn: MIRZA BK

PORTUGAL
Livraria Portugal
Rua do Carmo 70-74
Apart. 2681
1117 Lisboa Codex   Tel.: (01) 347.49.82/3/4/5
Telefax: (01) 347.02.64

SINGAPORE – SINGAPOUR
Information Publications Pte
Golden Wheel Bldg.
41, Kallang Pudding, #04-03
Singapore 1334 Tel. 741.5166
Telefax: 742.9356

SPAIN – ESPAGNE
Mundi-Prensa Libros S.A.
Castelló 37, Apartado 1223
Madrid 28001 Tel. (91) 431.33.99
Telefax: (91) 575.39.98

Libreria Internacional AEDOS
Consejo de Ciento 391
08009 – Barcelona Tel. (93) 488.34.92
Telefax: (93) 487.76.59
Llibreria de la Generalitat
Palau Moja
Rambla dels Estudis, 118
08002 – Barcelona
(Subscripcions) Tel. (93) 318.80.12
(Publicacions) Tel. (93) 302.67.23
Telefax: (93) 412.18.54

SRI LANKA
Centre for Policy Research
c/o Colombo Agencies Ltd.
No. 300-304, Galle Road
Colombo 3 Tel. (1) 574240, 573551-2
Telefax: (1) 575394, 510711

SWEDEN – SUÈDE
Fritzes Fackboksföretaget
Box 16356
Regeringsgatan 12
103 27 Stockholm Tel. (08) 23.89.00
Telefax: (08) 20.50.21
Subscription Agency-Agence d'abonnements
Wennergren-Williams AB
Nordenflychtsvägen 74
Box 30004
104 25 Stockholm Tel. (08) 13.67.00
Telefax: (08) 618.62.32

SWITZERLAND – SUISSE
Maditec S.A. (Books and Periodicals - Livres
et périodiques)
Chemin des Palettes 4
1020 Renens/Lausanne Tel. (021) 635.08.65
Telefax: (021) 635.07.80

Mail orders only - Commandes
par correspondance seulement
Librairie Payot
C.P. 3212
1002 Lausanne Telefax: (021) 311.13.92

Librairie Unilivres
6, rue de Candolle
1205 Genève Tel. (022) 320.26.23
Telefax: (022) 329.73.18

Subscription Agency - Agence d'abonnement
Naville S.A.
38 avenue Vibert
1227 Carouge Tél.: (022) 308.05.56/57
Telefax: (022) 308.05.88

See also – Voir aussi :
OECD Publications and Information Centre
Schedestrasse 7
D-W 5300 Bonn 1 (Germany)
Tel. (49.228) 21.60.45
Telefax: (49.228) 26.11.04

TAIWAN – FORMOSE
Good Faith Worldwide Int'l. Co. Ltd.
9th Floor, No. 118, Sec. 2
Chung Hsiao E. Road
Taipei Tel. (02) 391.7396/391.7397
Telefax: (02) 394.9176

THAILAND – THAÏLANDE
Suksit Siam Co. Ltd.
113, 115 Fuang Nakhon Rd.
Opp. Wat Rajbopith
Bangkok 10200 Tel. (662) 251.1630
Telefax: (662) 236.7783

TURKEY – TURQUIE
Kültur Yayinlari Is-Türk Ltd. Sti.
Atatürk Bulvari No. 191/Kat. 13
Kavaklidere/Ankara Tel. 428.11.40 Ext. 2458
Dolmabahce Cad. No. 29
Besiktas/Istanbul Tel. 160.71.88
Telex: 43482B

UNITED KINGDOM – ROYAUME-UNI
HMSO
Gen. enquiries Tel. (071) 873 0011
Postal orders only:
P.O. Box 276, London SW8 5DT
Personal Callers HMSO Bookshop
49 High Holborn, London WC1V 6HB
Telefax: (071) 873 8200
Branches at: Belfast, Birmingham, Bristol, Edin-
burgh, Manchester

UNITED STATES – ÉTATS-UNIS
OECD Publications and Information Centre
2001 L Street N.W., Suite 700
Washington, D.C. 20036-4910 Tel. (202) 785.6323
Telefax: (202) 785.0350

VENEZUELA
Libreria del Este
Avda F. Miranda 52, Aptdo. 60337
Edificio Galipán
Caracas 106 Tel. 951.1705/951.2307/951.1297
Telegram: Libreste Caracas

YUGOSLAVIA – YOUGOSLAVIE
Jugoslovenska Knjiga
Knez Mihajlova 2, P.O. Box 36
Beograd Tel. (011) 621.992
Telefax: (011) 625.970

Orders and inquiries from countries where Distribu-
tors have not yet been appointed should be sent to:
OECD Publications Service, 2 rue André-Pascal,
75775 Paris Cedex 16, France.

Les commandes provenant de pays où l'OCDE n'a
pas encore désigné de distributeur devraient être
adressées à : OCDE, Service des Publications,
2, rue André-Pascal, 75775 Paris Cedex 16, France.

Subscription to OECD periodicals may also be
placed through main subscription agencies.

Les abonnements aux publications périodiques de
l'OCDE peuvent être souscrits auprès des
principales agences d'abonnement.

OECD PUBLICATIONS, 2 rue André-Pascal, 75775 PARIS CEDEX 16
PRINTED IN FRANCE
(61 92 14-1) ISBN 92-64 13754-8 - N° 46243